An Exploration of Density, Realms, Dimensions, and the Multiverse

An Exploration of Density, Realms, Dimensions, and the Multiverse

Dakota Frandsen

To the student of Bald and Bonkers Network Academy

First Printing, 2025

To the student of Bald and Bonkers Network Academy

First Printing, 2025

CONTENTS

1

A Universe Beyond Our Senses

Our perception of reality is fundamentally shaped by our senses – sight, hearing, touch, taste, and smell. These five gateways to experience have served us well in navigating the everyday world, allowing us to interact with our environment and build a shared understanding of reality. However, the limitations of our sensory apparatus are profound, and recognizing these limitations is the first step in appreciating the vastness of the universe beyond our immediate grasp. Consider, for instance, the electromagnetic spectrum. Our eyes perceive only a tiny sliver of it – the visible light – while the vast majority, including radio waves, microwaves, infrared, ultraviolet, X-rays, and gamma rays, remain invisible to us. We are, in essence, blind to the vast majority of the electromagnetic energy that surrounds us, a reality that profoundly shapes our perception of the cosmos.

Similarly, our hearing is limited to a specific range of frequencies. Infrasound and ultrasound, while present in the environment, remain inaudible to us, effectively silencing a significant portion of the sonic world. The sense of touch, too, has its limitations. We cannot directly perceive subtle variations in pressure, temperature, or electromagnetic fields, all of which may hold information about the universe inaccessible to our tactile senses. Our senses, finely tuned as they are for survival on Earth, are poorly equipped to detect or interpret phenomena beyond the realm of our everyday experiences.

Throughout history, humankind's attempts to understand the universe have been consistently hampered by these perceptual limitations. Early civilizations, lacking the scientific tools and knowledge to probe the unseen, relied on mythology and folklore to explain the wonders and mysteries of the cosmos. The sun, moon, and stars were often attributed to divine beings or celestial forces, reflecting our limited ability to comprehend their true nature. Even with the advent of modern science, our understanding remains strikingly limited. The vastness of space, the intricacies of quantum mechanics, and the mysteries of dark matter and dark energy consistently challenge our current models and theories. The very act of observing a quantum system, for instance, alters its state, highlighting the inherent limitations of our attempts to objectively comprehend the universe. The Heisenberg uncertainty principle, a cornerstone of quantum mechanics, underscores this fundamental limit.

The exploration of the unknown, however, is not merely a scientific endeavor; it is an inherently human pursuit, deeply rooted in our curiosity and our desire to understand our place in the cosmos. From ancient astronomers charting the movements of stars to modern scientists probing the secrets of subatomic particles, our quest for knowledge has consistently pushed the boundaries of our understanding. This relentless pursuit, fueled by both intellectual excitement and a practical need to survive and thrive, has been a defining characteristic of the human species. It's the driving force behind our exploration of space, our attempts to decipher the human genome, and our quest to unlock the mysteries of consciousness.

Popular culture, in its own way, reflects this enduring fascination with the unknown. Science fiction stories, from H.G. Wells' "The War of the Worlds" to modern cinematic explorations of interstellar travel and alternate realities, grapple with the possibilities beyond our immediate experience. These narratives, often pushing the boundaries of scientific plausibility, speak to our inherent desire to venture into the uncharted territories of the universe. Similarly, the enduring popularity of paranormal investigations, ghost stories, and UFO sightings points to a deep-seated fascination with the inexplicable – a testament to our innate curiosity and our desire to understand phenomena that lie beyond the scope of our current scientific understanding.

The historical development of scientific instruments illustrates the expanding reach of human perception. The invention of the telescope, for instance, dramatically expanded our view of the cosmos, revealing celestial bodies and phenomena previously invisible to the naked eye. Similarly, the microscope opened up a whole new world of microscopic organisms and cellular structures, revealing the intricacies of life at the smallest scales. Modern scientific instruments, such as particle accelerators and radio telescopes, continue to push the boundaries of our observational capabilities, probing ever deeper into the fundamental structure of matter and the vastness of the universe.

However, even with these sophisticated instruments, our ability to perceive and understand the universe remains fundamentally limited. The sheer scale of the cosmos, encompassing billions of galaxies each containing billions of stars, stretches our comprehension to its limits. We are still grappling with fundamental questions about the nature of dark matter and dark energy, which constitute the majority of the universe's mass-energy content yet remain elusive to our direct observation. The mysteries of black holes, with their extreme gravity warping space and time, continue to challenge our understanding of the laws of physics.

Moreover, the very nature of scientific inquiry is an ongoing process of refinement and revision. Theories that are accepted as facts today may be overturned tomorrow in light of new evidence or a deeper understanding. This inherent uncertainty, while potentially frustrating, is also a source of excitement and intellectual stimulation. It motivates us to ask

more questions, to conduct more experiments, and to seek ever more profound answers. The history of science is littered with examples of previously held beliefs being replaced by more accurate and comprehensive models. The shift from a geocentric to a heliocentric model of the solar system, or the development of quantum mechanics that fundamentally altered our understanding of the microscopic world, serves as potent reminders of the dynamism of scientific progress and the ever-evolving nature of our understanding.

Therefore, acknowledging the inherent limits of our perception and the vastness of the universe beyond our immediate grasp is not a cause for despair, but rather a call to action. It encourages us to embrace the unknown with a sense of wonder and a commitment to rigorous scientific inquiry. The exploration of the unknown, be it through the lens of established scientific methodologies or through open-minded consideration of alternative perspectives, is not only intellectually stimulating, but also profoundly valuable. It holds the potential to unlock revolutionary breakthroughs in our understanding of the universe and our place within it. The journey into the unknown, while challenging and demanding, is ultimately a journey of discovery, a testament to the unyielding curiosity and persistent pursuit of knowledge that defines the human spirit. This understanding lays the foundation for the explorations of density, dimensions, and the multiverse that will follow in subsequent chapters.

Our exploration of the universe beyond our senses necessitates a deeper understanding of the fundamental properties of matter and energy. One such property, crucial to the discussion that follows, is density. Density, simply defined, is the mass of a substance per unit volume. This seemingly straightforward concept holds profound implications for our understanding of reality, particularly when considering the possibility of other realms or dimensions. We experience density in our everyday lives without often explicitly acknowledging it. The air we breathe, the water we drink, and the rock beneath our feet all possess different densities. Air, being relatively less dense, allows us to move through it with ease. Water, with a higher density, offers more resistance. Rock, significantly denser than both, presents a substantial barrier.

These everyday examples illustrate the profound influence of density on our interactions with the physical world. Imagine, however, extending this concept to the cosmological scale. Could variations in density, perhaps on scales far beyond our current observational capabilities, create distinct regions of spacetime with drastically different properties? Could pockets of extremely high density give rise to realms inaccessible to us, realms that might harbor forms of matter and energy unlike anything we've ever encountered? Conversely, regions of extremely low density could potentially create a void-like existence, dramatically different from our own.

This concept of density-defined realities is not merely a fanciful speculation. It finds resonance within established physical theories. Consider, for instance, the concept of a black hole. A black hole represents an extreme case of density, where matter has been compressed into an incredibly small volume, creating a gravitational field so intense that not even light can escape. The density at the singularity of a black hole is theoretically infinite, a concept that stretches our understanding of physics to its limits. The existence of black holes, confirmed through observational evidence, provides a compelling example of how extreme density can dramatically alter the fabric of spacetime.

Furthermore, the concept of a multiverse, a theoretical framework proposing the existence of multiple universes, often invokes the idea of differing physical laws and constants. These variations could be, at least in part, attributed to differing densities of matter and energy across these universes. A universe with a significantly higher density of matter than our own might experience fundamentally different physical phenomena, exhibiting vastly different laws of physics and supporting entirely novel forms of life or even matter states.

The idea that variations in density might create different planes of existence offers a potential explanation for some paranormal phenomena. For example, reports of ghosts or apparitions could be interpreted as interactions with entities inhabiting regions of spacetime with density significantly different from our own. These entities, operating under different physical laws, might be capable of interacting with our reality in unpredictable and seemingly inexplicable ways. The fleeting nature of such encounters could be explained by the inherent instability of these regions, or by the limitations of our sensory apparatus, which might not be equipped to directly perceive them.

Similarly, the persistent fascination with UFOs and extraterrestrial life could be understood within this framework. Perhaps some of these encounters involve beings originating from regions of spacetime with different density properties. Their advanced technology, seemingly defying our understanding of physics, could be a consequence of operating within a realm governed by physical laws influenced by the density of that region. Their ability to seemingly appear and disappear without a trace might be attributed to their ability to manipulate density or navigate between regions of varying density.

To better visualize this concept, consider an analogy: imagine a three-dimensional space filled with various substances of differing densities. Air, water, and rock coexist within this space, each occupying distinct volumes with distinct properties. Now, imagine extending this analogy to other dimensions, perhaps higher spatial dimensions that our senses cannot perceive directly. Within these higher dimensions, regions of varying density might exist, creating distinct planes of existence analogous to our three-dimensional space. These planes could interact with our own, albeit in limited and often imperceptible ways.

The study of dark matter and dark energy provides further credence to the idea of density influencing the nature of reality. Dark matter, constituting a substantial portion of the universe's mass, remains largely invisible to us, interacting with ordinary matter primarily through its gravitational effects. The nature of dark matter is still unknown, but its significant gravitational influence suggests a density far beyond our current comprehension. Dark energy, an even more mysterious entity, is thought to be responsible for the accelerating expansion of the universe. Its pervasive nature and influence imply a ubiquitous presence of energy with a unique density that shapes the evolution of the entire cosmos.

This leads us to ponder a critical question: how might we detect or even interact with these density-defined realms? The answer, undoubtedly, lies in developing scientific tools and methodologies far beyond our current capabilities. New instruments capable of detecting subtle variations in spacetime curvature, perhaps based on novel concepts in quantum gravity, might offer a path towards understanding and even interacting with regions of significantly different densities. Moreover, deeper theoretical understanding of the nature of dark matter and dark energy could unlock new insights into the dynamics of these density-defined realms.

The exploration of density as a defining factor in reality opens up a universe of possibilities, challenging our understanding of the cosmos and our place within it. It presents a compelling framework for exploring the seemingly inexplicable, offering a potential bridge between established scientific theories and the often-controversial realm of parapsychology. By acknowledging the limitations of our senses and embracing the vast unknown, we can embark on a journey of discovery that holds the potential to fundamentally reshape our understanding of the universe. Our sensory limitations, far from being a barrier, may serve as a pointer to a reality far more intricate and expansive than we can currently comprehend, a reality shaped, at its most fundamental level, by the seemingly simple property of density. The path forward requires a blend of rigorous scientific inquiry, open-mindedness, and a deep appreciation for the profound mysteries that still shroud the vast cosmos.

Our journey into the unexplored landscapes of reality leads us inevitably to the multiverse hypothesis – a concept that stretches the boundaries of our imagination and challenges the very foundations of our understanding of the cosmos. The idea that our universe might be just one among countless others, each with its own unique laws of physics, constants, and perhaps even forms of life, is both exhilarating and profoundly unsettling. But what is the scientific basis for such a bold assertion?

One of the most compelling theoretical frameworks supporting the multiverse hypothesis is string theory. String theory attempts to unify all fundamental forces of nature – gravity, electromagnetism, the strong nuclear force, and the weak nuclear force – within a single, elegant mathematical framework. Instead of treating fundamental particles as point-like objects, string theory postulates that they are tiny, vibrating strings of energy. The different vibrational modes of these strings correspond to different particles, and the properties of these particles are determined by the way the strings vibrate.

The mathematics of string theory, however, requires the existence of extra spatial dimensions beyond the three spatial dimensions we experience in everyday life. These extra dimensions are typically thought to be compactified or curled up at scales far smaller than anything we can currently probe experimentally. But the precise manner in which these extra dimensions are compactified leads to a vast landscape of possible universes, each with its own unique set of physical laws and constants. This vast landscape of possibilities is often referred to as the "string theory landscape," and it provides a strong theoretical argument for the existence of a multiverse.

The sheer number of possible universes predicted by string theory is staggering. Some estimates suggest that there could be 10-500 or even more possible universes, each with its own unique configuration of physical laws and constants. This vastness challenges our intuitive understanding of reality, pushing us to contemplate the possibility that our universe, with its specific set of physical laws and constants that permit the existence of stars, planets, and life as we know it, might be just one highly improbable outcome among a near-infinity of possibilities.

Another compelling line of reasoning supporting the multiverse hypothesis comes from inflationary cosmology. Inflationary cosmology suggests that the early universe underwent a period of extremely rapid expansion, far exceeding the expansion rate predicted by the standard Big Bang model. This period of inflation, driven by a hypothetical field known as the inflation field, smoothed out the initial inhomogeneities in the universe, resulting in the remarkably uniform distribution of matter and energy we observe today.

However, the inflationary scenario is not restricted to a single period of inflation. According to some inflationary models, the process of inflation could be eternal, leading to the continuous creation of new "bubble universes" within a larger, overarching "multiverse." Each of these bubble universes would have its own unique set of physical laws and constants, potentially differing significantly from our own. This process of continuous inflation and bubble universe creation presents a compelling mechanism for the emergence of a multiverse, albeit one that is inherently difficult to test experimentally.

The implications of the multiverse hypothesis are far-reaching and profound. If our universe is just one among countless others, it raises fundamental questions about our place in the cosmos, the uniqueness of our existence, and the very nature of reality itself. It challenges the anthropocentric view of the universe, suggesting that our specific physical laws and constants might not be unique or finely tuned, but rather just one outcome among many possible outcomes.

The multiverse hypothesis also has important implications for the fine-tuning problem – the observation that certain physical constants in our universe appear to be finely tuned for the emergence of life. If our universe is just one among many, the fine-tuning problem becomes less perplexing. In a vast multiverse, universes with physical constants unfavorable to life would simply not be observed by life, while universes with suitable conditions would naturally contain observers capable of pondering the fine-tuning problem. This is often referred to as the "multiverse solution" to the fine-tuning problem.

Despite the compelling theoretical arguments for the multiverse hypothesis, it remains a highly speculative concept. Currently, there is no direct observational evidence to support the existence of a multiverse. Testing the multiverse hypothesis directly presents a formidable challenge, as it requires venturing beyond the confines of our observable universe, something well beyond our current technological capabilities.

However, indirect evidence might potentially be found by searching for subtle signatures of other universes in the cosmic microwave background radiation or in other cosmological observations. Theoretical advancements in quantum gravity, particularly those that aim to unify quantum mechanics and general relativity, could also shed light on the feasibility and implications of the multiverse hypothesis.

The exploration of the multiverse hypothesis is not just a philosophical exercise; it represents a cutting-edge area of research in cosmology and theoretical physics. It pushes the boundaries of our scientific understanding, challenging us to reconsider our assumptions about the nature of reality and the universe. The quest to understand whether our universe is unique or just one of countless others is a grand scientific challenge, one that requires a synthesis of theoretical physics, observational cosmology, and a profound appreciation for the vastness and mystery of the cosmos.

The implications of a multiverse extend beyond the realm of theoretical physics. The possibility of other universes, each with its own physical laws and possibly even different forms of life, has captured the imagination of scientists, philosophers, and the public alike. It raises profound questions about the nature of reality, consciousness, and our place in the cosmos. Are the laws of physics truly fundamental, or are they merely local manifestations of a deeper, more universal structure? Could there be other forms of consciousness, perhaps radically different from our own, in these other universes?

Exploring these questions, even if we cannot directly prove the existence of other universes, pushes the boundaries of human understanding and inspires us to explore the limits of our knowledge. The search for evidence, whether direct or indirect, is a journey of discovery that promises to unveil profound insights into the structure and nature of our reality, challenging our perceptions and deepening our comprehension of the universe. The very act of engaging with such a grand concept expands our intellectual horizons and compels us to consider the unimaginable possibilities that lie beyond our current comprehension. The pursuit of understanding the multiverse, therefore, is not just a scientific endeavor, but a philosophical and even spiritual quest to grasp the full extent of existence itself.

Our exploration of a universe beyond our senses necessitates a careful examination of fundamental concepts: realms, dimensions, and the very nature of reality. These terms, often used interchangeably in popular discourse, require precise definition within a scientific framework, acknowledging the limitations of our current understanding and embracing the possibilities beyond the confines of our everyday experience.

The term "realm," in the context of this exploration, signifies a distinct region or plane of existence, potentially characterized by unique physical laws, energy densities, and perhaps even fundamental constants differing from our own. It's not necessarily a spatial separation in the way we conventionally understand distance, but rather a difference in the fundamental fabric of reality itself. One could envision realms existing concurrently within the same space-time continuum but interacting only under specific, perhaps yet-unknown, conditions. Think of it as different radio frequencies coexisting;

they occupy the same space but can only be accessed with the correct receiver. Similarly, different realms might exist simultaneously, but their interaction might depend on factors we haven't yet identified, possibly related to subtle energy shifts or quantum entanglement.

The concept of "density" plays a crucial role in differentiating realms. Varying energy densities could influence the physical laws operating within a given realm. A realm with significantly lower energy density might exhibit different physical properties, leading to fundamentally different interactions between particles and forces. For example, a realm with a much lower density could support forms of matter and energy that are impossible within our own high-density realm. This could offer a compelling explanation for the alleged encounters with entities that seem to defy our understanding of physics, such as ghosts or spirits. These entities, if they exist, might inhabit realms with different energy densities, allowing them to interact with our own in limited ways, perhaps manifesting as fleeting disturbances or subtle changes in energy fields.

Furthermore, the possibility of variations in fundamental constants across realms cannot be overlooked. The values of constants such as the speed of light, the gravitational constant, and the Planck constant, which are fundamental to our understanding of the universe, may differ significantly in other realms. These variations could lead to dramatically different physical laws, resulting in realms with radically different structures and properties. In some realms, for instance, gravity might be significantly weaker, allowing for structures and life forms unimaginable within our own universe. Conversely, realms with extremely high gravitational forces could lead to completely different physical phenomena, potentially collapsing all matter into highly dense, singularity-like states.

The definition of "dimensions" requires a nuanced approach, moving beyond the conventional three spatial dimensions and one time dimension that form the familiar fabric of our spacetime. In physics, dimensions refer to independent parameters required to specify a point in space or spacetime. We experience three spatial dimensions—length, width, and height—but string theory and other theoretical frameworks posit the existence of additional spatial dimensions, potentially curled up or compactified at scales too small to be directly observed with current technology. These extra dimensions could be crucial in understanding the nature of realms. Each realm could represent a different configuration of these extra dimensions, resulting in different physical laws and characteristics.

The concept of the multiverse, discussed previously, strongly supports the idea of multiple realms. In the multiverse scenario, our universe is but one among countless others, each potentially possessing its own unique set of physical laws, fundamental constants, and even dimensionality. The differences between these universes or realms could stem from different initial conditions in the Big Bang, or perhaps from the branching of universes through quantum processes. The concept of "bubble universes," generated during inflationary epochs, postulates a multiverse where each universe expands independently, creating realms that are fundamentally separate yet coexist within a larger framework.

Considering the nature of reality itself opens up yet another layer of complexity. Our understanding of reality is fundamentally shaped by our sensory perception and the tools we use to observe the universe. However, our senses and instruments may only perceive a limited portion of the total reality. What we perceive as "reality" might just be a specific manifestation or a particular "slice" of a far more extensive reality encompassing multiple realms and dimensions. Quantum mechanics, with its concept of superposition and entanglement, adds another layer of complexity, suggesting that reality might be far more fluid and less deterministic than our classical models would suggest. The act of observation, according to the Copenhagen interpretation, influences the outcome, implying a degree of interconnectedness and interdependence that defies our conventional notion of causality.

To fully grasp the potential existence of realms and dimensions beyond our current perception, we must be open to considering interpretations of reality that extend beyond our intuitive understanding. The notion of a holographic universe, for instance, proposes that the entire three-dimensional universe is encoded on a two-dimensional surface at its boundary. This would imply that our perception of a three-dimensional world is merely a projection, a hologram of a deeper, more fundamental reality. Similarly, quantum physics suggests that reality exists as a probabilistic superposition of states, collapsing into a single outcome only when measured. These considerations force us to reconsider the fundamental assumptions we make about the structure and nature of reality itself, opening the door for the possibility of multiple interwoven realities existing simultaneously.

The concept of parallel universes, often featured in science fiction, represents another fascinating avenue of exploration. The many-worlds interpretation of quantum mechanics, while controversial, postulates the existence of multiple universes, each representing a different possible outcome of a quantum event. Every time a quantum measurement is made, the universe branches, creating a new universe where a different outcome occurs. In this scenario, countless parallel universes exist, each representing a slightly different path that reality could have taken, encompassing a multitude of realms and experiences beyond our own. Although this remains a highly speculative concept, the exploration of its theoretical implications serves as a powerful tool to expand our conceptual understanding of reality's vastness.

Ultimately, defining realms, dimensions, and the nature of reality requires a multidisciplinary approach, integrating insights from cosmology, theoretical physics, parapsychology, and even philosophy. While direct observation of other realms remains beyond our current technological capabilities, the development of advanced theoretical models, coupled with innovative experimental approaches, may pave the way towards a deeper understanding of the vast expanse of realities that may lie beyond our senses. The journey itself, an intellectual exploration into the unknown, is as important as the destination; the very act of questioning and exploring the possibilities expands our knowledge and challenges us to reassess our place within the cosmos. The pursuit of understanding the nature of reality, whether it involves rigorous scientific inquiry or philosophical speculation, leads to a richer, more complete, and profoundly awe-inspiring view of the universe and our place within its complex tapestry.

The exploration of realms beyond our immediate sensory perception necessitates a rigorous and unbiased approach. While the very notion of "paranormal" suggests phenomena outside the established framework of scientific understanding, dismissing these claims outright without careful investigation would be a disservice to the pursuit of knowledge. The scientific method, with its emphasis on observation, hypothesis formation, experimentation, and analysis, remains the cornerstone of our understanding of the universe. However, applying this method to the study of phenomena often deemed paranormal requires a nuanced approach, acknowledging the inherent challenges and limitations.

One of the primary difficulties lies in the nature of the evidence itself. Many purported paranormal events rely heavily on anecdotal accounts, eyewitness testimonies, and subjective experiences. While such accounts can be intriguing and even suggestive, they lack the reproducibility and quantifiable data that are essential for rigorous scientific investigation. Confirmation bias, the tendency to interpret information to confirm pre-existing beliefs, can significantly affect the reliability of anecdotal evidence. Witnesses may unconsciously exaggerate or misinterpret events, or their accounts may be influenced by suggestion or social pressures. Therefore, differentiating between genuinely unexplained phenomena and those attributable to misperception, misinterpretation, or outright fabrication is crucial.

This doesn't imply that anecdotal evidence should be completely disregarded. Instead, it should be treated with a healthy dose of skepticism and viewed as a potential starting point for further investigation. Intriguing anecdotal accounts can highlight patterns or suggest areas warranting more rigorous scrutiny. For instance, numerous reports of purported poltergeist activity in specific locations might motivate a scientific team to conduct a thorough investigation, employing objective instruments to measure physical phenomena such as electromagnetic fields, temperature fluctuations, or acoustic anomalies. The presence of unusual patterns or correlations in such measurements could then be used to formulate testable hypotheses.

Rigorous scientific investigation of paranormal claims often involves employing advanced instrumentation and statistical analysis to rule out alternative explanations. For example, studies investigating purported psychokinesis, the alleged ability to influence physical systems with the mind, may use sophisticated equipment to measure subtle changes in physical systems, eliminating any potential influence from known physical forces or mechanisms. This rigorous approach helps ensure that any observed effects are unlikely to be due to chance or experimental error. Similarly, studies of alleged telepathy, the supposed transmission of thoughts or information between minds, might employ double-blind experimental designs to mitigate bias and ensure the objectivity of the results.

The exploration of purported precognitive abilities, the supposed ability to foresee future events, presents a unique challenge to the scientific method. The difficulty lies in the inherent difficulty of testing and validating precognitive claims. A prediction that comes true might be a matter of chance or coincidence rather than a manifestation of genuine precognitive ability. Scientific studies attempt to overcome this challenge by employing rigorous statistical analysis, including the use of control groups and objective scoring methods. They might compare the accuracy of precognitive predictions against what one would expect through pure chance. Statistically significant deviations from chance expectations could lend weight to the claims, but such evidence is typically subject to intense scrutiny and often requires replication by independent researchers to be widely accepted.

Furthermore, the challenge in studying paranormal phenomena lies in the absence of a universally accepted theoretical framework. Unlike established branches of physics, where well-established theories provide a foundation for investigation and prediction, the field of parapsychology lacks a unifying theoretical model. This lack of theoretical grounding makes it difficult to formulate testable hypotheses and interpret experimental results. Developing a comprehensive theoretical framework that can explain paranormal phenomena within a scientific context is a significant hurdle that must be overcome before these phenomena can be fully integrated into mainstream scientific discourse.

Nevertheless, several researchers and institutions have dedicated themselves to the rigorous scientific study of these phenomena. The Rhine Research Center, for example, has a long history of investigating parapsychological claims using controlled experiments. Their research, while often met with skepticism from the broader scientific community, represents a serious attempt to apply scientific methodology to a controversial field. Similarly, other researchers worldwide are engaged in exploring related areas, using sophisticated techniques to gather and analyze data.

The history of science is replete with examples of phenomena initially considered paranormal or supernatural that subsequently found explanations within established scientific frameworks. The aurora borealis, once attributed to celestial beings or divine intervention, is now understood as a consequence of charged particles interacting with the Earth's atmosphere. The ability to cure certain diseases through seemingly miraculous methods might be attributed to the advancement of medical science that was previously unavailable. Such historical examples highlight the importance of maintaining an open mind while adhering to the principles of rigorous scientific investigation. It is possible that some phenomena currently labeled as paranormal might eventually be explained through a more advanced understanding of physics, biology, or the human mind.

In conclusion, applying the scientific method to the exploration of the paranormal requires a nuanced approach that balances rigorous skepticism with open-mindedness. While anecdotal evidence should be treated with caution, it can serve as a starting point for further investigation. The use of controlled experiments, advanced instrumentation, and robust statistical analysis are crucial in distinguishing between genuine unexplained phenomena and those attributable to other factors. The development of a robust theoretical framework is crucial for advancing understanding in this field. The ongoing scientific investigation of these phenomena, despite the inherent challenges, is essential for advancing our understanding of the universe and our place within it. The ultimate goal is not to prove or disprove the existence of the paranormal, but rather to employ the scientific method to explore these phenomena rigorously and objectively, whatever the outcome might be. This approach allows us to expand the boundaries of scientific knowledge and deepen our understanding of reality, even if that reality extends beyond the confines of our current comprehension.

2

Exploring Density and its Effects on Reality

The concept of density, often overlooked in everyday life, plays a pivotal role in shaping the physical universe as we know it. Density, simply defined as mass per unit volume, is not merely a static property; it's a dynamic factor that dictates the behavior and interaction of matter at all scales, from the subatomic to the cosmic. Understanding density's influence is crucial to unraveling the complexities of reality, especially when considering the possibility of realms beyond our immediate experience.

Let's begin with the familiar states of matter: solid, liquid, gas, and plasma. Each state is defined by the density and arrangement of its constituent particles. Solids, characterized by their rigid structure, possess high density due to the tightly packed arrangement of their atoms or molecules. The strong intermolecular forces hold these particles in fixed positions, resulting in a relatively incompressible material. Consider a block of iron: its high density arises from the closely packed iron atoms, each contributing to its overall mass within a relatively small volume. A similar mass of, say, styrofoam, occupies a far larger volume due to the less dense arrangement of its constituent polymers, and the presence of air pockets further reduces its density. This difference in density dramatically alters their properties—strength, flexibility, and ability to float, for example.

Liquids, less dense than solids, exhibit a more fluid structure. Their particles are still close together, but their movement is not restricted to fixed positions. The intermolecular forces are weaker than in solids, allowing for greater freedom of motion and enabling liquids to conform to the shape of their container. Water, for example, has a density significantly lower than ice, resulting in ice floating on water, a crucial factor in aquatic ecosystems. The transition between the solid and liquid state, known as melting, involves overcoming the intermolecular forces holding the solid's structure together. The energy required to accomplish this is directly related to the strength of the intermolecular forces and, consequently, the density difference between the two phases.

Gases exhibit even lower densities than liquids. The particles in a gas are widely dispersed and move relatively independently, only interacting briefly during collisions. The intermolecular forces are extremely weak, allowing gases to expand freely to fill any available volume. The density of a gas is highly dependent on pressure and temperature; increasing the pressure forces the particles closer together, increasing density, while increasing temperature increases the kinetic energy of the particles, causing them to move further apart and thus decreasing density. This behavior is described by the ideal gas law, which relates pressure, volume, temperature, and the number of particles in a gas. The difference in density between a gas and its liquid or solid counterparts can be enormous, underscoring the significant role of interparticle forces and packing efficiency in determining density.

Plasma, often considered the fourth state of matter, is characterized by extremely high temperatures that ionize the constituent atoms, resulting in a mixture of free-moving ions and electrons. Plasma exhibits remarkably high energy states and low density compared to solids and liquids but can reach densities similar to gases at extremely high pressures. The behavior of plasma is governed by electromagnetic forces, rather than just the intermolecular forces that dominate the behavior of solids, liquids, and gases. Examples of plasma include stars, lightning, and fluorescent lights. The extreme temperatures and energies associated with plasma lead to unique properties not observed in other states of matter. For instance, plasma can conduct electricity, emit light, and be influenced significantly by magnetic fields – features that have a wide range of applications, from fusion reactors to plasma displays.

The transitions between these states of matter – melting, freezing, boiling, condensation, sublimation, and deposition – are fascinating examples of density changes in action. These transitions are not abrupt but rather involve a gradual change in density as the temperature and pressure of a substance vary. Consider water: as ice melts, its density increases until it reaches a maximum at approximately 4°C. This anomaly is crucial for aquatic life, as it ensures that the densest water is found at the bottom of lakes and rivers, preventing complete freezing and preserving life below the surface. Furthermore, the dramatic density difference between water in its liquid and gaseous phases is responsible for the evaporation process, whereby water molecules with sufficient kinetic energy escape the liquid phase, transitioning to a gaseous state with far lower density.

These phase transitions highlight the sensitivity of density to external factors such as temperature and pressure. Changes in density don't simply affect the physical properties of a substance; they can fundamentally alter its behavior. This principle takes on added significance when we consider the speculative possibility of different realms or dimensions. Could variations in density, perhaps at a quantum level or within other dimensional spaces, create realms with drastically different physical laws and properties?

One intriguing possibility is that the density of a region of spacetime could influence the perceived properties of that region. A region of exceptionally high density, for example, might manifest as a seemingly impenetrable barrier, while a region of extremely low density could exhibit properties akin to a void or even a different plane of existence. This idea might potentially explain anecdotal accounts of paranormal phenomena such as alleged encounters with ghosts or extraterrestrial beings. Perhaps these encounters represent interactions with regions of spacetime possessing density variations that create conditions unrecognizable within our ordinary sensory perceptions. The difficulty, of course, lies in detecting and measuring such density variations, which may exist beyond the capabilities of our current technology.

Furthermore, the concept of density extends beyond the macroscopic realm into the quantum world. Quantum field theory describes the universe as a complex interplay of quantum fields, each permeating all of spacetime. The density of these quantum fields, representing the intensity of their influence at a given point, plays a significant role in determining the physical properties of that region. It is conceivable, though highly speculative, that significant variations in quantum field densities might lead to regions of spacetime exhibiting properties that defy our current understanding of physics.

The question of whether different densities of quantum fields or variations in spacetime curvature, which is directly related to mass-energy density according to Einstein's theory of general relativity, could create separate realms or dimensions remains open. However, the exploration of this possibility is crucial in expanding our understanding of the universe and its potential complexity. While our current scientific methods may not suffice to investigate the reality of other dimensions or realms, the conceptual framework based on density variations provides a potential starting point for further exploration, connecting the apparently disparate fields of parapsychology and cosmology.

The journey into the unknown always requires a blend of rigorous scientific inquiry and a willingness to consider possibilities that lie beyond the currently established framework of knowledge. The exploration of density, from the well-established properties of matter in its various states to the speculative implications of density variations in spacetime and quantum fields, represents a vital step in that journey. It is through such interdisciplinary explorations that we might one day unravel the mysteries of reality and potentially glimpse the existence of dimensions and realms beyond our immediate perception. The challenge remains formidable, but the potential rewards—a deeper understanding of the universe and our place within it—are immeasurable. The ongoing scientific investigation, combined with a thoughtful consideration of potential implications, offers the best path forward in this exploration of the unknown.

Building upon the established relationship between density and the observable states of matter, we can now venture into more speculative territory. The hypothesis that significant density variations could generate distinct realms or planes of existence is a bold one, yet it warrants careful consideration. This hypothesis offers a potential framework for interpreting phenomena that currently defy conventional scientific explanations, such as purported encounters with ghosts or reports of interdimensional travel.

The concept of density, as we've seen, isn't merely a static property of matter; it's a dynamic variable influenced by factors such as temperature, pressure, and the fundamental forces governing the universe. In the macroscopic world, these variations lead to phase transitions between solids, liquids, gases, and plasma. But what if density variations were not limited to these familiar states? What if variations in density at the quantum level, or even at the level of spacetime itself, could create regions with radically different physical laws and properties?

Einstein's theory of general relativity provides a crucial link between density and the curvature of spacetime. Mass-energy density dictates the curvature of spacetime, and this curvature, in turn, influences the motion of objects within that spacetime. A region of high mass-energy density will curve spacetime significantly, leading to strong gravitational effects. Conversely, a region of low mass-energy density will have a relatively flat spacetime. Could extreme variations in mass-energy density, beyond what we observe in the known universe, create regions of spacetime with properties so different from our own as to constitute separate realms or dimensions?

Imagine a region of spacetime with an exceptionally high density, perhaps exceeding the density found within neutron stars or black holes. Such a region might be characterized by extraordinarily strong gravitational forces, effectively creating a barrier that is essentially impenetrable to ordinary matter. It might appear to us as a void, a space where nothing can exist or pass through, though other beings or entities existing within this higher density realm might perceive it quite differently. They might find their own form of existence perfectly adapted to these extreme conditions, and their experience of this region would be entirely different from ours.

Conversely, a region of extremely low density could potentially behave like a void, but a very different kind of void than the one described above. Here, the lack of significant mass-energy might result in a spacetime with minimal curvature, exhibiting properties entirely foreign to our experience. For example, the very fabric of space and time might behave in unexpected ways; causality might be different, or even nonexistent. This is highly speculative, but such a region might represent a bridge to other dimensions or serve as a gateway to entirely different universes.

The potential for different density regions to create distinct planes of existence also opens up possibilities within the context of quantum physics. Quantum field theory describes the universe as a collection of interacting quantum fields that pervade all of spacetime. The density of these fields, representing the strength of their influence at a given point, is a dynamic variable that influences the physical properties of that point. Perhaps significant variations in quantum field densities could give rise to regions with drastically altered physical laws, forming separate realms with unique properties and physical constants.

Consider the possibility of quantum entanglement extending across these hypothetical density-defined realms. Entangled particles, even if separated by vast distances or existing in different realms due to density variations, would still exhibit correlated behavior. This could provide a means of communication or interaction between these different planes of existence, although the mechanisms of such interaction remain highly speculative. It might imply that information, in some form, could be transferred across the density gradients separating these realms.

The challenge, of course, lies in our limited ability to detect and measure such density variations. Our current scientific instruments are primarily designed to investigate the universe within the range of densities we experience directly. Detecting significant density variations, especially at the quantum level or across hypothetical dimensional boundaries, requires new theoretical frameworks and experimental techniques. Nevertheless, the search is paramount. This exploration may require a radical rethinking of our observational tools and measurement strategies.

It's important to note that this hypothesis is not intended to offer definitive answers but rather to stimulate further exploration and discussion. The idea that density variations could lead to different realms or dimensions is speculative, and its validation requires considerable further research. However, the potential explanatory power of this concept in regard to unexplained phenomena, such as alleged paranormal encounters, warrants its serious consideration.

It is conceivable that alleged interactions with ghosts or otherworldly beings could be interpreted as encounters with regions of spacetime or quantum field densities that lie beyond our normal sensory perception and measurement capabilities. These encounters, therefore, wouldn't represent a violation of physical laws but rather an interaction with a part of the universe governed by different physical laws, determined by unique density parameters.

This requires a departure from purely materialistic explanations. If different realms are governed by different laws of physics, this suggests a more complex, multi-layered reality than the simplistic models we often employ. This multi-layered view doesn't necessarily discount materialism, but rather expands and enriches it, allowing for the possibility of interactions between different levels of reality.

The connection between density variations and the potential existence of other realms offers a tantalizing bridge between seemingly disparate fields of study, particularly parapsychology and cosmology. By bringing these fields into dialogue, we might be able to develop a more comprehensive understanding of the universe and our place within it. It might also shed light on long-standing mysteries that have baffled scientists and philosophers for centuries.

The road ahead is long and arduous. Our current understanding of the universe is incomplete, and many open questions remain. However, by approaching the subject with a blend of scientific rigor and open-mindedness, we can hope to make progress in understanding the true nature of reality. The exploration of density, its variations, and its potential implications for the existence of other realms offers a compelling path towards that goal. The universe, it seems, may be far more expansive and wondrous than we currently conceive. The key lies in pushing the boundaries of our scientific understanding and embracing the unknown. The potential for discovery is immense, and the journey promises to be both challenging and profoundly rewarding.

Building upon the concept of density-defined realms, we can now explore the intriguing possibility of interdimensional travel or interaction facilitated by these density gradients. This is a highly speculative area, venturing far beyond the currently established boundaries of physics, but it offers a compelling avenue for theoretical exploration. The very notion of "interdimensional travel" evokes images from science fiction, but if we approach it through the lens of density variations, we might find a framework that aligns more readily with known scientific principles, albeit in a highly extrapolated way.

One potential mechanism for interdimensional travel, inspired by Einstein's theory of general relativity, involves the concept of wormholes. These are hypothetical tunnels through spacetime, connecting two distant points in the universe, or even two different universes. The formation of a wormhole is thought to require exotic matter with negative mass-energy density, a concept still largely theoretical. However, if we consider regions of extremely low density, as discussed previously, we might find conditions that could, at least theoretically, approximate the requirements for wormhole formation. The low density region, essentially a "thinning" of spacetime, could act as a conduit connecting our reality to another, a realm defined by a different density profile.

The challenge, of course, lies in the extreme conditions necessary for wormhole formation. The gravitational forces at play would be immense, likely exceeding anything we have directly observed. Moreover, the stability of a wormhole is a significant concern; the slightest disruption could cause it to collapse, trapping anything that had ventured inside. However, the existence of density gradients, which could potentially create regions of differing spacetime curvature, suggests a mechanism by which such an unstable wormhole might be temporarily stabilized or even naturally maintained, at least within the context of our hypothetical density-defined realms.

Another potential mechanism for interdimensional interaction hinges on the concept of quantum entanglement. As previously mentioned, entangled particles remain correlated even when separated by vast distances. This correlation extends beyond the constraints of classical physics, hinting at a deeper connection between these particles than our current understanding allows. If we extend this principle to the hypothetical density-defined realms, we could imagine entangled particles existing in different realms, maintaining their correlation despite the separation.

This correlation could provide a channel for communication or even for the transfer of information or energy between these realms. Imagine two entangled particles, one residing within our reality and the other within a higher-density realm. Any manipulation of one particle might instantaneously affect the other, potentially allowing for a form of interdimensional communication. This, however, relies on several significant leaps of faith. It presupposes that entanglement persists across the boundaries of density-defined realms, which is currently purely speculative. The information transfer mechanism itself remains a significant theoretical challenge. The nature of the information exchanged would also need further exploration. Would it be limited to quantum information, or could more complex, classical information be transmitted?

The difficulty in exploring these possibilities lies in our current limitations in detecting and measuring the density variations responsible for creating these hypothetical realms. Our existing scientific instruments are not equipped to detect such subtle density fluctuations, especially at the quantum level or across potential dimensional boundaries. Therefore, new theoretical frameworks and experimental techniques are necessary to even begin testing these ideas. This could involve the development of sensors far beyond our current capabilities, perhaps sensors that can probe the quantum vacuum or measure spacetime curvature with unprecedented precision.

The development of such tools and theories requires a multidisciplinary approach, bringing together physicists, mathematicians, and even parapsychologists. The latter, though often met with skepticism within mainstream science, offer anecdotal evidence and empirical observations of phenomena that might be interpreted within the framework of interdimensional interaction. While these observations are often not considered scientifically rigorous, their potential to shed light on these questions should not be ignored completely. A careful, critical examination of these accounts, alongside rigorous theoretical and experimental work, could potentially unlock important insights.

The role of consciousness in interdimensional interaction is also a fascinating, albeit controversial, area of exploration. Some theories suggest that consciousness itself plays a significant role in shaping reality, and that conscious intention could play a crucial role in facilitating interdimensional travel or interaction. This is a complex issue, fraught with philosophical and metaphysical implications, but its potential relevance cannot be dismissed.

Consider the possibility of conscious intention influencing the quantum field densities that might define different realms. If consciousness can affect quantum processes, as some theories suggest, then it might be possible to manipulate density gradients through concentrated mental effort, thus creating or accessing pathways between realms. This aligns with some interpretations of altered states of consciousness, such as deep meditation or near-death experiences, which some report as involving experiences that could be interpreted as interdimensional encounters.

The exploration of these possibilities requires a shift in perspective, a move beyond strictly materialistic explanations of reality. A more holistic approach, incorporating elements of parapsychology and quantum consciousness alongside established physics, might be necessary to make progress in this field. This does not imply a rejection of scientific rigor but rather a broadening of the scope of inquiry. A willingness to consider unconventional ideas, alongside rigorous testing and analysis, is critical.

Furthermore, considering the vastness of the universe and the limitations of our current understanding, it is not unreasonable to posit the existence of other dimensions or realms governed by different physical laws, accessible through density gradients. Our current understanding of physics is based on observations within a specific range of conditions. Extrapolating those observations to extreme conditions – extremely high or low densities – may lead to predictions that defy our intuition, but that does not necessarily mean they are incorrect. Indeed, such predictions might offer profound insights into the true nature of reality.

The challenges are immense, but the potential rewards are equally significant. A deeper understanding of density gradients and their implications could lead to breakthroughs in physics, cosmology, and even our understanding of consciousness. It could revolutionize our view of the universe, revealing a far more complex and interconnected reality than we currently imagine. The journey into this uncharted territory is daunting, but the potential for discovery is immense, and the pursuit of knowledge, in its broadest sense, is an endeavor worth undertaking. The exploration of density and its effects on reality, including the possibility of interdimensional travel, represents a significant frontier in our quest to unravel the mysteries of the cosmos.

The previous discussion of density gradients and their potential role in shaping reality naturally leads us to consider the fundamental connection between energy and density. Energy, in its various forms, is inextricably linked to the fabric of spacetime, influencing its curvature and potentially creating the density variations that could give rise to separate realms or dimensions. A deeper understanding of this relationship is crucial to further explore the possibility of interdimensional interaction. Einstein's famous equation, $E=mc^2$, elegantly demonstrates the equivalence of energy and mass. Mass, in turn, contributes directly to the gravitational field, a key factor in determining spacetime curvature. Therefore, regions of high energy density are also regions of high mass density, and thus experience stronger gravitational effects.

Conversely, regions of low energy density correspond to regions of low mass density and weaker gravitational effects. This simple relationship has profound implications for the concept of density-defined realms.

Consider, for instance, the immense energy density found at the heart of a black hole. The gravitational force in this region is so extreme that it warps spacetime to such a degree that even light cannot escape. This extreme warping of spacetime, a direct consequence of the incredibly high energy density, could potentially create a bridge or a conduit to another realm, another dimension defined by a drastically different density profile. While the specifics remain highly theoretical, the principle suggests that regions of extremely high energy density might serve as portals to other realities.

Conversely, regions of extremely low energy density could also potentially lead to unusual effects. The quantum vacuum, for example, is often described as a state of lowest possible energy, yet it is far from empty. Instead, it is a seething cauldron of virtual particles constantly popping into and out of existence. These fluctuations in the quantum vacuum, while minute, represent energy density variations at a fundamental level. Could these fluctuations be responsible for creating subtle shifts in spacetime curvature, leading to the formation of micro-dimensional variations or "quantum wormholes"? Such microscopic variations may not be traversable by macroscopic objects, but they could still play a significant role in quantum entanglement and information transfer across hypothetical density-defined realms.

High-energy phenomena in the universe, such as supernovae and gamma-ray bursts, also offer compelling examples of how energy distribution can impact density. These cataclysmic events release unimaginable quantities of energy into their surroundings, temporarily creating regions of extremely high energy density. This drastic increase in energy density could, in principle, cause significant distortions in spacetime, potentially creating temporary or even persistent interdimensional connections. The extremely high energies involved might even lead to the formation of exotic particles or states of matter not yet observed in our reality, further blurring the lines between our perceived reality and other potential realms.

The concept of dark energy, a mysterious force driving the accelerated expansion of the universe, further complicates this picture. Dark energy is thought to have a negative pressure and a uniform density, influencing the universe's expansion on a cosmological scale. While its nature remains unknown, its pervasive influence throughout the cosmos hints at a subtle yet powerful influence on spacetime curvature, contributing to the overall density profile of the universe and potentially influencing the creation and stability of density-defined realms.

Exploring the role of energy in defining density requires examining high-energy physics in detail. The Standard Model of particle physics describes the fundamental building blocks of matter and their interactions, but it leaves many questions unanswered. For instance, the nature of dark matter, a mysterious substance making up a significant portion of the universe's mass, remains a puzzle. Understanding dark matter's properties and its distribution could provide critical insights into the overall energy density distribution of the universe and its role in shaping the fabric of reality.

Similarly, the search for new physics beyond the Standard Model is crucial. Theories such as supersymmetry and string theory posit the existence of extra dimensions and exotic particles, further complicating but also enriching our understanding of the universe's energy landscape. These theoretical frameworks, though still under development, offer the potential to provide a more complete picture of how energy contributes to density variations and the formation of potential realms.

Moreover, the relationship between energy and density isn't merely static; it's dynamic. The universe is constantly evolving, with energy being created, transformed, and redistributed across vast cosmic scales. This dynamic interplay of energy and density suggests that the boundaries between realms are not necessarily fixed but might fluctuate or even shift over time. The density gradients that define these realms might ebb and flow, creating temporary windows of interdimensional interaction or communication.

This dynamic aspect underscores the need for a temporal perspective when studying these phenomena. A snapshot in time might only reveal a limited picture, while a long-term analysis could reveal fluctuating density gradients and patterns of energy distribution that could be crucial to understanding interdimensional connections. Cosmology, with its focus on the evolution of the universe, thus plays a critical role in unraveling these mysteries.

The exploration of these complex relationships requires advanced theoretical and experimental approaches. Developing more sophisticated models of spacetime, incorporating quantum field theory and gravitational effects, is essential. The development of highly sensitive detectors that can measure minute variations in energy density, potentially at the quantum level, would be revolutionary. This would allow us to directly probe the fabric of spacetime and potentially detect subtle fluctuations that could indicate the presence of interdimensional connections or variations in the density of reality.

Furthermore, the development of novel computational techniques, specifically designed to simulate the complex interactions of energy and matter on a cosmological scale, would be crucial in testing and validating theoretical models. Such simulations could potentially reveal subtle patterns and fluctuations in energy density that are beyond our current observational capabilities. A multidisciplinary approach, bringing together theoretical physicists, cosmologists, and computer scientists, is essential for tackling this daunting task.

It's important to reiterate that much of this discussion remains speculative. However, the connection between energy and density is a powerful concept that offers a plausible framework for considering the existence of other realms or dimensions. It provides a unifying theme that could connect seemingly disparate areas of physics, cosmology, and even parapsychology. By systematically exploring this relationship, we might be able to shed light on some of the most profound mysteries of existence, ultimately expanding our understanding of reality itself beyond our current limited perceptions. The journey, filled with challenges and unanswered questions, is nonetheless an exciting and intellectually stimulating quest, promising untold rewards for those who dare to explore the unknown.

Building upon the established link between energy and density, we can now delve into how these variations might directly influence the perception of reality itself. The idea is not simply that different densities exist, but that these differences fundamentally alter how an observer within that density experiences the universe. Consider the analogy of a fish in a dense, underwater environment versus a bird soaring through the relatively less dense atmosphere. Their perceptions of the world, their physical capabilities, and even their fundamental understanding of "reality" are vastly different, shaped entirely by the density of their respective environments.

This concept extends far beyond simple differences in medium. We can imagine vastly different densities of spacetime itself, influencing not only the physical laws but also the very nature of perception within a given realm. A realm with extremely high energy density, for instance, might experience time dilation, causing events to unfold at vastly different rates compared to our own reality. Observers within such a high-density realm could perceive time as slowed, stretched, or even non-linear, leading to a radically different experience of causality and temporal perception. Similarly, spatial dimensions could also be distorted, resulting in a warped perception of distances and geometries.

Imagine a realm where the density of spacetime is so high that the very fabric of space is compressed, making it feel "smaller" to an inhabitant, despite its objective physical size. Conversely, a realm with extremely low spacetime density could appear vast and expansive, even infinite, regardless of its actual dimensions. This effect could create a profound illusion, leading inhabitants of such realms to perceive themselves as inhabiting an incredibly vast or confined universe, depending entirely on the density of their environment. Such variations in the perception of space and time could fundamentally alter the very nature of reality as perceived by their inhabitants. These beings might possess vastly different concepts of distance, duration, and even simultaneity compared to our own.

Furthermore, the density of a realm could directly influence the types of matter and energy that can exist within it. High-density realms might favor the formation of exotic particles and states of matter not found in our reality. Conversely, low-density realms might be incapable of supporting complex structures, or even matter as we understand it, resulting in a reality fundamentally different from anything we can readily comprehend. This suggests that different realms, defined by varying densities, might not only have different physical laws but also host fundamentally different types of life, forms of consciousness, and methods of interaction.

This concept leads us to consider the possibility that what we perceive as "paranormal" phenomena might be interactions, albeit indirect and limited, with other realms of varying densities. For example, the perception of ghosts, often described as ethereal or translucent, could be explained by a difference in density between our reality and the realm from which these entities originate. If these "ghosts" exist within a lower density realm, they might interact with our reality only weakly, creating faint impressions or residual effects that are perceived as paranormal activity. Their low density could explain their seemingly insubstantial nature and ability to pass through solid objects.

Similarly, alleged extraterrestrial encounters could be explained by the inhabitants of a realm with a different energy density than ours. Such beings might use energy sources or methods of propulsion that are completely unknown to us, making their interactions appear inexplicable or even supernatural. Their interaction with our reality, mediated by the difference in density, could result in fleeting encounters or technologically advanced phenomena that defy our current understanding of physics. Their very appearance and capabilities might be radically different, shaped by the unique physical laws of their native realm.

The perceptual shifts caused by density variations are not necessarily limited to dramatic differences. Subtle fluctuations in energy density within our own realm could also produce subtle but noticeable effects. Consider the possibility that variations in the density of quantum fluctuations, or even fluctuations in the density of dark energy, could momentarily alter the perception of reality at a macroscopic level. Such small, localized variations might manifest as glitches in perception, subtle shifts in perception of space and time, or even inexplicable experiences that cannot be explained by our current understanding of physics. These occurrences, while potentially rare, could help to explain some otherwise inexplicable phenomena that are often dismissed or ignored.

Another crucial aspect to consider is the influence of the observer themselves. An observer's own density, or their relative position within a density gradient, could determine how they perceive reality. An observer positioned within a high-density region might experience the universe differently than an observer positioned within a lower-density region. The experience of reality is not a passive reception of data but an active process of interpretation, shaped by the characteristics of both the observer and the observed. This suggests that the very act of observation might influence the density and thus the perception of reality itself, leading to a feedback loop between observer and environment.

Furthermore, the concept of density and its impact on the perception of reality has implications for understanding consciousness and its relationship to the universe. It suggests that consciousness, rather than being confined to a particular location or medium, might be a property of the universe itself, existing across varying densities of energy. This panpsychist view would suggest that consciousness is a fundamental aspect of reality, intrinsic to the fabric of spacetime and present in varying degrees across different realms and dimensions.

This perspective offers a fascinating new lens through which to view the relationship between consciousness and the cosmos, suggesting that our experience of reality is shaped by the complex interplay of energy, density, and the very nature of consciousness itself. The density of our environment, and the density of our own consciousness, may determine how we perceive the universe and our place within it.

The exploration of density and its effect on reality is a far-reaching and complex undertaking. It demands a multidisciplinary approach, integrating concepts from physics, cosmology, parapsychology, and even philosophy. However, by engaging with these seemingly disparate fields, we can begin to construct a more comprehensive and accurate picture of the universe, one that acknowledges the potential for realities beyond our current grasp. The journey continues, with many mysteries still to unravel, but the path we are exploring promises to reveal profound truths about the nature of existence itself. The investigation into density and its effect on perception is not simply a scientific pursuit, but a philosophical journey into the very heart of what it means to exist and to perceive. The implications of understanding how density shapes reality are far-reaching, affecting our understanding of everything from the smallest quantum events to the vastness of the cosmos and the nature of consciousness itself. The work ahead is challenging, requiring innovative thinking and interdisciplinary collaboration, but the potential rewards—a deepened understanding of reality and our place within it—are immeasurable.

3

Dimensions Beyond Our Three

Our everyday experience is firmly rooted in three spatial dimensions: length, width, and height. We navigate the world using these coordinates, perceiving objects and events within this familiar framework. Yet, the theoretical landscape of physics suggests that this seemingly self-evident reality might be a significant simplification, a mere glimpse into a far more complex and multifaceted universe. The concept of dimensions beyond our three-dimensional perception is a cornerstone of many advanced physical theories, notably string theory and M-theory, which posit the existence of extra spatial dimensions curled up or compactified at scales far smaller than anything we can currently observe.

To understand the implications of these higher dimensions, we need to step back and examine the mathematical underpinnings of spatial dimensionality. At its core, the concept of a dimension is a way to specify the location of a point in space. In one dimension, a single number—for example, a position along a line—is sufficient to define a point's location. In two dimensions, like the surface of a plane, we need two numbers (coordinates) to uniquely identify a point, such as its x and y coordinates. Similarly, in our three-dimensional world, three coordinates—x, y, and z—are required. The number of coordinates needed to define a point's location corresponds to the number of spatial dimensions.

Extending this logic to higher dimensions is straightforward mathematically, although visualizing them becomes increasingly challenging. A four-dimensional space requires four coordinates, a five-dimensional space requires five, and so on. While we can't directly perceive these extra dimensions, their existence is supported by theoretical frameworks that elegantly unify different aspects of fundamental physics, such as gravity and quantum mechanics. The mathematical formalism allows for the consistent description of these higher dimensions, even if their physical manifestation remains elusive.

One common approach to visualizing higher dimensions uses analogies. Consider the analogy of a creature confined to a two-dimensional plane, like an ant on a sheet of paper. The ant can only move forward, backward, left, and right. It cannot comprehend the concept of "up" or "down," as these directions lie outside its two-dimensional experience. Similarly, we, confined to our three-dimensional world, might be unable to perceive the higher dimensions that might exist beyond our immediate perception. Just as the ant's world is a "slice" of a three-dimensional reality, our three-dimensional world could be a "slice" of a higher-dimensional reality.

Another helpful visualization involves the "shadow" analogy. Imagine a three-dimensional cube casting a shadow onto a two-dimensional plane. The shadow would appear as a complex, changing shape depending on the cube's orientation relative to the plane. An observer confined to the two-dimensional plane could not directly perceive the three-dimensional nature of the cube but could only infer its presence and structure from its varying shadow. Similarly, the effects of higher dimensions might be manifested in our three-dimensional universe as seemingly inexplicable phenomena, hints of a reality beyond our direct perception.

These higher dimensions are not necessarily vast, boundless spaces like our own three spatial dimensions. In many theoretical models, particularly string theory, these extra dimensions are "compactified," meaning they are curled up into tiny, subatomic structures. This compactification explains why we haven't directly observed them. Their size is so incredibly small, far smaller than the Planck length—the smallest scale meaningful in quantum mechanics—that they are effectively invisible to our current observational capabilities. This doesn't mean they are not there; it simply means that their effects manifest indirectly, through the interactions of particles and forces in our three-dimensional world.

The mathematical description of these compact dimensions often involves complex geometrical structures, such as Calabi-Yau manifolds. These manifolds are higher-dimensional spaces with intricate shapes and topologies that profoundly influence the physical laws governing our universe. The specific shape and configuration of these compactified dimensions determine the constants of nature, the properties of fundamental particles, and the overall structure of the universe. Different configurations lead to different universes with different physical laws, opening the possibility of a multiverse, where each universe is characterized by a unique set of compactified dimensions.

The implications of higher spatial dimensions extend to our understanding of gravity. General relativity, our current best description of gravity, is remarkably successful in describing the gravitational interactions on macroscopic scales. However, it faces challenges when trying to reconcile it with quantum mechanics. String theory and M-theory attempt to overcome this challenge by incorporating higher spatial dimensions. In these theories, gravity is not a fundamental force mediated by a particle, like the other forces, but rather a manifestation of higher-dimensional geometry. Gravitons, the hypothetical particles of gravity, could propagate and interact in these higher dimensions, offering a potential explanation for the weakness of gravity compared to the other fundamental forces.

Furthermore, the existence of higher spatial dimensions can also have profound consequences for our understanding of cosmology and the evolution of the universe. The early universe, in its incredibly dense and hot state, might have had different physical properties due to the interplay of higher dimensions. The phase transitions and symmetry breakings that shaped the universe's evolution could be deeply affected by the influence of these extra dimensions, possibly leading to vastly different outcomes than what we observe today.

Considering the potential existence of higher spatial dimensions challenges our fundamental assumptions about the nature of reality. It prompts us to reconsider our anthropocentric view of the universe, recognizing that our three-dimensional perception might be a limited perspective of a far richer and more complex reality. The journey into these higher dimensional realms is not just a mathematical exercise but a profound exploration of the very fabric of existence, expanding our understanding of the cosmos and our place within it. The pursuit of this understanding requires creative mathematical tools and innovative experimental approaches that push the boundaries of what we currently believe is possible, continuously revising our understanding of the universe and its hidden dimensions. This exploration requires interdisciplinary collaboration, bridging the gap between theoretical physics, mathematics, and other relevant fields, to create a holistic perspective that transcends traditional disciplinary boundaries. The quest to understand the spatial dimensions of our universe is a testament to the enduring human curiosity and our relentless pursuit of knowledge, constantly revealing deeper layers of reality that challenge our assumptions and expand our understanding of the cosmos.

Our exploration of dimensions beyond our familiar three has so far focused on spatial dimensions. However, the concept of dimensionality extends beyond mere spatial extent; time itself can be considered a dimension, albeit a dimension of a different character than the spatial ones. This is a pivotal concept, profoundly altering our understanding of the universe and our place within it. While we intuitively perceive time as a linear progression, a relentless march from past to future, the incorporation of time as a fourth dimension, within the framework of spacetime, revolutionizes this perspective.

Einstein's theory of special relativity, a cornerstone of modern physics, fundamentally alters our comprehension of time. It reveals that time is not absolute, an independent entity ticking uniformly for all observers regardless of their motion. Instead, relativity demonstrates that time is relative, its passage intertwined with the observer's motion and the gravitational field they experience. This means that the measurement of time intervals can vary depending on the observer's velocity relative to another observer. Two observers in relative motion will disagree on the duration of events.

Consider the classic example of the "twin paradox." Imagine a pair of identical twins. One twin embarks on a high-speed journey into space, while the other remains on Earth.

When the spacefaring twin returns, they will have aged less than their Earthbound sibling. This isn't a matter of physiological differences, but rather a consequence of the time dilation predicted by special relativity. The faster the twin travels, the slower their clock runs relative to the stationary twin's clock. This effect, although minuscule at everyday speeds, becomes significant at velocities approaching the speed of light.

This relativity of time arises from the fundamental interconnectedness of space and time, unified within the concept of spacetime. Spacetime is a four-dimensional continuum, where three spatial dimensions (length, width, height) are combined with a temporal dimension (time). Events are not simply located in space but are situated within spacetime, having both spatial and temporal coordinates. This unification is not merely a mathematical construct but a reflection of the profound interplay between space and time as revealed by Einstein's theory.

The implications of this interconnectedness are far-reaching. The geometry of spacetime determines the behavior of objects under the influence of gravity. In general relativity, Einstein extended special relativity to incorporate gravity, describing it not as a force but as a curvature of spacetime. Massive objects warp the fabric of spacetime, causing other objects to move along curved trajectories. This curvature manifests as the gravitational force we experience.

This curvature of spacetime further affects the passage of time. Time slows down in stronger gravitational fields. A clock located closer to a massive object, such as a black hole, will run slower than a clock located farther away. This phenomenon, known as gravitational time dilation, has been experimentally verified with remarkable accuracy using atomic clocks placed at different altitudes. The higher the altitude, the faster the clock runs, albeit by a minuscule amount.

The concept of spacetime as a unified four-dimensional continuum dramatically changes how we perceive the universe. Instead of viewing space and time as separate entities, we must consider them as intertwined aspects of a single framework. This holistic view provides a more accurate and complete description of the universe's dynamics. It is within this spacetime that the universe's events unfold, its structure takes shape, and its evolution proceeds.

The incorporation of time as a dimension also leads to intriguing possibilities regarding the nature of reality. Some interpretations of quantum mechanics suggest that time might be emergent rather than fundamental, arising from the interactions of microscopic constituents of the universe. If time is emergent, then its linear progression, as we perceive it, might only be an approximate description valid at macroscopic scales. At the Planck scale, where quantum gravitational effects are dominant, time's very nature could be radically different from our everyday experience.

Furthermore, the inclusion of time as a dimension opens the door to speculative possibilities concerning time travel. While currently outside the realm of practical possibility, the theoretical framework of spacetime allows for the conceptual exploration of scenarios in which the past, present, and future could be interconnected, potentially enabling movements between different points in time. This remains a subject of intense debate and exploration, involving discussions of paradoxes and potential causal violations.

Moreover, the combination of spatial and temporal dimensions provides a richer understanding of the universe's evolution. Cosmology, the study of the universe's origin, structure, and evolution, relies heavily on the spacetime framework. The expansion of the universe, a cornerstone of modern cosmology, is described within the four-dimensional spacetime, where distances between galaxies increase with time. The Big Bang, the universe's beginning, is not merely a point in time but a point in spacetime, marking the singularity from which the universe emerged.

The implications of time as a dimension extend far beyond the realm of theoretical physics. They affect our understanding of causality, the relationship between cause and effect. In our everyday experience, cause precedes effect, but in a four-dimensional spacetime, this linear ordering might be an artifact of our limited perception. In principle, events occurring at different times and locations could be causally connected in more complex ways than we intuitively imagine.

The exploration of time as a dimension is an ongoing journey, pushing the boundaries of our knowledge and challenging our fundamental assumptions about the universe. It requires sophisticated mathematical tools and innovative theoretical models, such as string theory and loop quantum gravity, which aim to unify general relativity and quantum mechanics within a framework that incorporates both spatial and temporal dimensions in a consistent way. The unification of these theories, however, remains one of the grand challenges of theoretical physics. This journey into the depths of spacetime necessitates interdisciplinary collaboration between physicists, mathematicians, and philosophers, creating a collective exploration into the fabric of reality. The investigation continues to unveil deeper layers of understanding, continuously challenging our perceptions and expanding our comprehension of the universe's enigmatic depths. The quest to fully understand the nature of time as a dimension represents a testament to humanity's unrelenting pursuit of knowledge, constantly refining our understanding of the cosmos and our place within its intricate tapestry of spacetime.

Our journey into dimensions beyond the familiar three has led us to consider time as a fourth dimension, interwoven with space to form the fabric of spacetime. But the story doesn't end there. The possibility of additional spatial dimensions, beyond the three we experience directly, is a recurring theme in advanced theoretical physics. These extra dimensions, if they exist, are not simply extensions of the length, width, and height we are accustomed to; they are of a fundamentally different character, curled up, compactified, or hidden from our direct perception.

One of the most promising frameworks for exploring these higher spatial dimensions is string theory. This elegant theoretical model posits that the fundamental building blocks of the universe are not point-like particles, as in the standard model of particle physics, but rather tiny, vibrating strings. The different vibrational modes of these strings correspond to different particles and forces. Crucially, string theory requires extra spatial dimensions to be mathematically consistent. The number of these extra dimensions varies depending on the specific version of the theory, but most versions require at least six or seven additional dimensions beyond the three we readily experience.

These extra dimensions in string theory are not the vast, expansive spaces we might imagine. Instead, they are thought to be "compactified," meaning they are curled up into incredibly tiny, subatomic structures, too small to be detected with our current technology. Imagine a garden hose: from far away, it appears one-dimensional, a simple line. But up close, you see it has a second dimension, its circumference. Similarly, the extra dimensions in string theory might be curled up to such a small size that they are effectively invisible at macroscopic scales.

The compactification of these extra dimensions is a crucial aspect of string theory. The specific way these dimensions are curled up dictates the physical laws and constants we observe in our universe. Different compactifications lead to different universes, each with its own unique set of physical characteristics. This notion introduces the concept of a "landscape" of possible universes, each corresponding to a different configuration of the extra dimensions. This landscape of possibilities is vast, perhaps containing a virtually infinite number of universes, each with potentially unique physical laws and characteristics. Our universe, with its specific set of constants and laws, would then be just one realization within this vast multiverse.

String theory's mathematical elegance and its capacity to incorporate gravity alongside quantum mechanics make it a highly compelling candidate for a theory of everything – a unified description of all the fundamental forces and matter in the universe. However, it remains a work in progress, facing significant challenges in terms of experimental verification. The scales at which these extra dimensions are hypothesized to exist are far beyond our current experimental reach. Developing techniques to probe these dimensions directly presents one of the greatest challenges in modern physics.

Extending from string theory is M-theory, a more comprehensive theoretical framework that encompasses string theory as a special case. M-theory proposes the existence of even more dimensions, generally eleven, with the extra dimensions having properties far beyond simple compactification. In M-theory, these additional dimensions play a more dynamic role, shaping the properties of spacetime and influencing the interactions between fundamental particles. M-theory is even more mathematically complex than string theory, making its testability and empirical validation a significant hurdle.

The implications of higher-dimensional physics extend far beyond the purely theoretical realm. If these extra dimensions exist, they could have profound implications for our understanding of gravity, cosmology, and even paranormal phenomena. The weakness of gravity compared to other fundamental forces, for example, might be explained by the "leakage" of gravitational force into these extra dimensions. Gravity's influence might be diluted because it spreads out not just through our three familiar spatial dimensions, but also into these hidden dimensions, making it appear weaker in our three-dimensional perspective.

Moreover, the existence of extra dimensions could have profound implications for cosmology. The early universe, during the Big Bang, may have had a significantly different topology and geometry due to the influence of these extra dimensions. Understanding how these dimensions evolved and interacted with the three spatial dimensions we experience could provide crucial insights into the universe's earliest moments and its subsequent evolution. The expansion of the universe, for instance, might be influenced by the dynamics of these extra dimensions.

Beyond these conventional implications, the existence of higher dimensions opens up fascinating avenues for speculative explorations. Some physicists suggest that these extra dimensions could provide a framework for understanding phenomena traditionally considered paranormal. Interactions with these dimensions could potentially explain observations that seem to defy conventional physics, such as alleged encounters with ghosts or other unexplained occurrences. This speculative realm, however, requires careful consideration and a balance between open-mindedness and rigorous scientific scrutiny. While intriguing, these connections remain highly speculative and require further investigation and development within a robust scientific framework.

The pursuit of understanding higher dimensions remains a profound and challenging endeavor, pushing the limits of human ingenuity and mathematical sophistication. While experimental verification presents a formidable challenge, the theoretical frameworks provided by string theory and M-theory offer potential explanations for a variety of phenomena, bridging the gap between seemingly disparate aspects of physics. The journey into these higher dimensions is not just about expanding our knowledge of the universe, but about fundamentally altering our perception of reality itself, challenging our assumptions about space, time, and the very nature of existence. As our understanding progresses, the potential implications will continue to unfold, promising to revolutionize our understanding of the cosmos and our place within it. The possibilities are vast, and the exploration continues. The pursuit of understanding these higher dimensions is a testament to humanity's ongoing quest to unravel the mysteries of the universe, forever pushing the boundaries of our knowledge and understanding. The quest is far from over; the adventure continues.

Our exploration of dimensions beyond the familiar three has touched upon the theoretical frameworks of string theory and M-theory, suggesting the existence of compactified or dynamically interacting extra spatial dimensions. While these theories provide a mathematical framework, the question naturally arises: how might we, as beings confined to our three-dimensional perception, possibly access or interact with these higher dimensions? This question takes us into the realm of highly speculative physics, but one that offers fascinating possibilities.

One of the most intriguing concepts in this context is that of wormholes. Predicted by Einstein's theory of general relativity, wormholes are hypothetical tunnels through spacetime, potentially connecting distant regions of the universe, or even different universes entirely. These tunnels are not simple holes in space; they are hypothesized to be bridges linking separate points, potentially allowing for traversal between vastly separated locations or even different temporal periods. The geometry of a wormhole is complex and often depicted as a throat connecting two "mouths" embedded in spacetime. The stability of such a wormhole presents a significant challenge, as they are predicted to be extremely unstable, collapsing almost immediately unless exotic matter with negative mass-energy density is used to hold them open. The existence of such exotic matter remains entirely hypothetical, presenting a significant hurdle in the possibility of wormhole traversal.

However, the theoretical possibility of wormholes remains a powerful catalyst for imagining interactions with higher dimensions. Some theoretical models propose that these higher dimensions might be connected to our own through wormhole-like structures. Navigating such a connection would require not only understanding the geometry of the wormhole but also overcoming the immense gravitational forces involved. Moreover, the nature of the higher-dimensional space on the other side of the wormhole is unknown, and traversing it could lead to unpredictable consequences, including potentially harmful interactions with unknown physics. The challenges involved are staggering, but the sheer audacity of the concept drives continued theoretical research.

Another avenue to consider is the fascinating phenomenon of quantum entanglement. This counterintuitive quantum effect involves two or more particles becoming linked in such a way that their fates are intertwined, regardless of the distance separating them. Measuring the state of one entangled particle instantaneously determines the state of the other, even if they are light-years apart. This instantaneous correlation has led to much speculation about its potential implications for faster-than-light communication or even travel.

While entanglement doesn't directly suggest a method for accessing higher dimensions in the same way that wormholes might, it raises questions about the nature of space and time at the quantum level. It's possible that the entanglement phenomenon, at its deepest level, could be linked to the interconnectedness of our three-dimensional world with higher dimensions. If we could harness and understand entanglement completely, perhaps we could use it as a tool to probe these other realms. However, this is highly speculative.

The use of quantum technologies, particularly advanced quantum computers, could potentially offer new avenues for exploring these concepts. Quantum computers exploit quantum mechanics to perform calculations far beyond the capabilities of classical computers. Their ability to harness entanglement could provide a significant advantage in modeling and simulating the complex physics of higher dimensions and phenomena like wormholes. By developing sufficiently powerful quantum simulations, we might be able to glean insights into the dynamics of higher-dimensional spaces and the possibility of interaction, even without the ability to physically traverse them. This theoretical approach provides a more accessible path to exploring the possibilities compared to directly attempting to manipulate exotic matter or create wormholes.

Beyond wormholes and entanglement, we must also consider the possibility of altering the density of spacetime itself. As previously discussed, density variations could create different planes or realms of existence. Perhaps by manipulating the density of our spacetime region, we could somehow "tune" it to resonate with or interact with a higher-dimensional space. However, the methods of manipulating spacetime density at this level remain entirely within the realm of science fiction. Current scientific understanding does not provide any mechanism for such manipulation, and any attempt to do so could have catastrophic consequences.

It's important to emphasize that all the methods discussed for accessing higher dimensions remain firmly in the realm of theoretical speculation. There is no current empirical evidence to support the existence of wormholes that are traversable, or of ways to manipulate spacetime density or entanglement to achieve this goal. However, the theoretical possibility, rooted in fundamental physics, is profoundly stimulating. The challenges involved are immense, requiring breakthroughs in our understanding of quantum gravity, exotic matter, and the fundamental nature of spacetime itself. This pursuit, however challenging, is intrinsically linked to our quest for a more complete understanding of the universe and our place within it.

Furthermore, the ethical considerations of interacting with higher dimensions, if such interaction becomes possible, are profound. What are the potential risks involved in opening up a wormhole or manipulating spacetime density? Could we unintentionally unleash catastrophic forces or encounter beings or phenomena beyond our comprehension? These are questions that must be seriously addressed before any attempts at accessing higher dimensions are even considered. A responsible approach necessitates a thorough understanding of the potential consequences, demanding an advanced level of scientific knowledge and ethical reflection. This careful consideration is paramount.

The exploration of methods for accessing higher dimensions is deeply interwoven with the development of advanced technologies. The creation of powerful quantum computers, more sophisticated detectors to detect gravitational waves, and potentially entirely new types of sensors could all play a crucial role. These technological advancements, in turn, drive the theoretical work, fostering a cyclical relationship between theory and experiment. The future exploration of higher dimensions will undoubtedly be characterized by this reciprocal relationship.

In summary, while the prospect of accessing higher dimensions currently resides in the realm of theoretical physics and science fiction, the inherent scientific plausibility, grounded in theories like string theory, M-theory, and general relativity, provides fertile ground for continued research and exploration. The potential implications are staggering, ranging from revolutionizing our understanding of the universe to raising profound ethical questions about the nature of reality and our responsibility in its exploration. The quest remains a daunting but undeniably fascinating challenge, pushing the boundaries of human ingenuity and imagination as we strive to unveil the deeper mysteries of the cosmos. The journey continues, and the possibilities, while speculative, remain profoundly intriguing. The potential rewards for uncovering even a glimpse of these higher dimensions would reshape our understanding of the universe in ways we can only begin to imagine. The effort, therefore, is worth pursuing, even if the path forward remains shrouded in significant unknowns and considerable challenges. The endeavor underscores the boundless curiosity and relentless pursuit of knowledge that define the human spirit.

Having explored potential avenues for accessing higher dimensions through theoretical physics, let us now consider a more speculative, yet potentially insightful, connection: the relationship between these higher dimensions and paranormal phenomena. The seemingly inexplicable nature of many paranormal events—ghost sightings, unexplained disappearances, telepathy, precognition, and other anomalous occurrences—has long defied conventional scientific explanation. However, the possibility that these phenomena arise from interactions with entities or energies residing in higher dimensions presents a compelling alternative framework.

One approach is to consider the concept of "density" as it relates to different dimensions. We have previously discussed how variations in spacetime density might create distinct realms or planes of existence. It is conceivable that higher-dimensional spaces possess significantly different densities compared to our own, rendering them invisible or undetectable using our current scientific instruments. Yet, under specific circumstances, these higher-dimensional entities or energies might manifest in our reality, leading to the observed paranormal effects. For example, a momentary fluctuation in spacetime density could allow a higher-dimensional entity to briefly interact with our three-dimensional space, resulting in a ghost-like apparition. The spectral nature of such sightings might reflect the partial or incomplete interaction with our reality, a glimpse through a veil of different dimensional densities. This model suggests that paranormal events aren't violations of physical laws, but rather subtle interactions governed by physics we have yet to fully grasp.

Another intriguing aspect is the possibility of higher-dimensional beings possessing unique capabilities not constrained by our three-dimensional laws of physics. If such beings exist, they could potentially manipulate spacetime, energy, or information in ways that would seem miraculous or paranormal to us. Telepathy, for example, might be a form of higher-dimensional communication that bypasses our conventional sensory limitations. Precognition could represent an ability to perceive temporal events across different dimensional planes, where the temporal constraints of our universe are less stringent. Similarly, psychokinesis, the purported ability to move objects with the mind, might involve the interaction of a higher-dimensional consciousness with our physical world through means that defy our current understanding of causality. These concepts require a radical shift in perspective, moving beyond our anthropocentric view of the universe to acknowledge the potential existence of other forms of intelligence and interaction.

The concept of a multiverse, consisting of numerous universes with varying physical laws and constants, further complicates the picture. Some theories propose that our universe might be embedded within a higher-dimensional space, interacting with other universes or parallel realities. This framework provides a theoretical context for the existence of otherworldly entities, dimensions inaccessible to us under normal circumstances. Paranormal events, then, could be interpreted as brief "leaks" or breaches between these parallel realities or universes, causing unusual phenomena in our own. The nature of these interactions, however, remains entirely speculative. Perhaps the conditions that give rise to these breaches are extremely rare or highly specific, explaining their infrequent occurrence. The exploration of such possibilities requires the integration of cutting-edge cosmology and theoretical physics with a careful, critical examination of paranormal phenomena. It necessitates a paradigm shift away from simply dismissing these events as impossible, towards an open-minded exploration of the theoretical possibilities.

It's crucial to acknowledge the challenges involved in scientifically investigating paranormal phenomena. The inherent unpredictability and lack of consistent experimental reproducibility pose major hurdles. Unlike the controlled environment of a physics laboratory, the study of paranormal events often relies on anecdotal evidence, making rigorous scientific analysis extremely difficult. However, advances in technology and data analysis techniques, coupled with the development of more sensitive detection instruments, could help alleviate some of these challenges. For example, advanced sensor arrays capable of detecting subtle variations in electromagnetic fields or gravitational waves could potentially capture elusive evidence of higher-dimensional interactions. Similarly, the development of more sophisticated data analysis algorithms might allow us to identify patterns and correlations in seemingly random paranormal events, hinting at underlying mechanisms.

Quantum entanglement, as previously discussed, could play a crucial role in understanding potential higher-dimensional interactions. The instantaneous correlation between entangled particles suggests a deeper connectivity than our three-dimensional perception allows. It's conceivable that this entanglement extends beyond our universe, linking it to higher dimensions and other universes, potentially providing a mechanism for paranormal phenomena. The challenges lie in understanding the nature of this entanglement on a cosmological scale and identifying how it could give rise to the observed paranormal effects. The very nature of entanglement defies classical intuition, suggesting the necessity of incorporating quantum mechanical considerations into our models of paranormal phenomena.

Furthermore, the investigation requires a multidisciplinary approach. Collaboration between physicists, psychologists, statisticians, and parapsychologists is crucial to develop rigorous methodologies for data collection and analysis. The potential biases and subjective interpretations that frequently plague the study of paranormal events need to be addressed through the development of standardized protocols and objective measurements. Only then can we move beyond speculative hypotheses and begin to develop testable models that could offer genuine scientific insights.

The concept of higher dimensions and their potential connection to paranormal phenomena remains firmly in the realm of speculation. However, exploring these possibilities can enrich our understanding of both the theoretical underpinnings of the cosmos and the nature of reality itself. The pursuit of a more complete understanding of the universe necessitates the inclusion of all available data, including that which currently falls outside of our conventional scientific framework.

The exploration of these speculative ideas must be approached with a balance of scientific rigor and intellectual humility. While embracing the possibility of higher-dimensional interactions as explanations for paranormal events offers an intriguing theoretical framework, it is essential to acknowledge the limitations of our current understanding. The lack of empirical evidence and the inherent challenges in scientifically investigating these phenomena underscore the need for continued, careful research.

The journey into understanding the cosmos and our place within it is a long and winding road. However, the pursuit of knowledge, driven by curiosity and the relentless questioning of what we believe we know, is central to the human endeavor. As we explore the potential connections between higher dimensions and paranormal phenomena, we must embrace both the excitement of discovery and the humility of acknowledging the vast unknowns that still lie ahead. The journey, even fraught with challenges and uncertainty, remains a profoundly important and rewarding endeavor. The integration of these seemingly disparate fields of study – theoretical physics, cosmology, and parapsychology – offers the potential for groundbreaking discoveries that could fundamentally alter our understanding of reality and our place within the vast expanse of the universe. The quest is ongoing, and the potential rewards, while uncertain, are immense.

4

The Multiverse: A Landscape of Realities

Having established a potential link between higher dimensions and paranormal phenomena, we now turn our attention to a concept that significantly expands the scope of this exploration: the multiverse. The idea of a multiverse, a collection of potentially infinite universes, each with its own unique physical laws and constants, provides a vast and fertile ground for speculation concerning the nature of reality and the possibility of other forms of existence. Several compelling models of the multiverse exist, each offering a distinct perspective on the structure and composition of reality beyond our own observable universe.

One of the most widely discussed models is the many-worlds interpretation (MWI) of quantum mechanics. Unlike the Copenhagen interpretation, which suggests that quantum wave functions collapse upon observation, MWI postulates that every quantum measurement causes the universe to split into multiple branches, each representing a different possible outcome. In this framework, every quantum event—from the decay of a radioactive atom to the interaction of particles in a collision—generates a branching of the universe, creating a vast, ever-expanding network of parallel universes. These parallel universes are not merely hypothetical constructs; they are considered equally real and physically existent as our own universe.

The implications of MWI are profound. It suggests that every decision we make, every chance event that unfolds, creates a new universe where a different outcome occurs. Imagine flipping a coin. In our universe, it lands on heads. But according to MWI, a parallel universe simultaneously exists where the coin landed on tails. This branching continues ad infinitum, with each quantum event generating further splits, leading to an almost incomprehensibly vast number of parallel realities. These realities may differ subtly, or they may be radically different, perhaps with different physical constants or even completely different laws of physics. This suggests that not only are there universes where seemingly minor events have different outcomes, but universes where the very fabric of reality is fundamentally altered.

The concept of a multiverse arising from MWI provides a fascinating context for paranormal phenomena. For instance, the experiences commonly described as déjà vu could potentially be interpreted as brief glimpses or "cross-talk" between parallel universes. The feeling of having experienced a situation before, even though it's objectively a novel experience, might be explained by a temporary overlap or interaction with a parallel universe where a similar event occurred. Similarly, precognitive experiences—feelings of knowing future events—could be explained as information leaking across the boundaries of these parallel realities. This is a purely speculative model, but it offers an intriguing theoretical framework for understanding experiences that seem to defy conventional explanations.

Another significant model of the multiverse arises from inflationary cosmology. Inflationary theory suggests that the universe underwent a period of extremely rapid expansion in its very early stages. While this theory is primarily used to explain the observed uniformity of the cosmic microwave background radiation, some models predict that this inflationary process might have been far more extensive than initially thought, potentially generating multiple "bubble universes" within a larger, encompassing multiverse. Each bubble universe would have its own set of physical laws and constants, potentially leading to vastly different realities.

The inflationary multiverse model is supported by the fact that the values of various physical constants within our universe appear remarkably fine-tuned for the emergence of life. This "fine-tuning" has led some physicists to speculate that our universe might be just one of many, with the majority being unsuitable for life. In this context, our universe is merely a fortunate accident, one that emerged with the specific constants that allowed for the development of stars, galaxies, and eventually, life itself. This vast landscape of universes, each with its own set of initial conditions and physical laws, is a powerful demonstration of the potential for a multiverse.

String theory, a theoretical framework that attempts to unify all fundamental forces of nature, also provides support for a multiverse. String theory postulates that the fundamental constituents of the universe are not point-like particles, but rather one-dimensional vibrating strings. The way these strings vibrate determines the properties of the particles and forces they represent. However, string theory permits a vast number of possible solutions, each representing a different universe with different physical laws and constants. This "landscape" of string theory solutions could give rise to a multiverse with an unimaginable number of universes. The sheer number of possible solutions makes it difficult to test string theory experimentally, but it remains one of the most intriguing theoretical frameworks for understanding the potential existence of a multiverse.

The concept of a multiverse raises many philosophical and scientific questions. If countless universes exist, how can we ever hope to observe or interact with them? What are the implications for our understanding of our place in the cosmos? These questions highlight the profound implications of the multiverse hypothesis, challenging our understanding of reality itself. The limitations of our current scientific instruments and theoretical frameworks make direct observation of other universes extremely challenging, if not impossible.

However, subtle observational effects, such as variations in the cosmic microwave background radiation or unexpected anomalies in the gravitational field, might potentially offer indirect evidence of the existence of other universes. The investigation into these possibilities remains a fertile area of scientific inquiry.

Furthermore, the idea of a multiverse has significant implications for the nature of consciousness and the anthropic principle. The anthropic principle states that the observed properties of the universe must be compatible with the existence of observers. If our universe is just one of many, then the observed "fine-tuning" of the physical constants might be explained by the simple fact that we could only exist in a universe that allowed for our emergence. In a multiverse with an infinite number of universes, some would inevitably have the necessary conditions for life to develop. This doesn't necessarily explain why our universe is the way it is, but it does offer a framework for understanding the conditions that allow for our existence.

The implications of the multiverse for paranormal phenomena are also profound. If countless universes exist, each with its own set of physical laws and constants, the possibility of interactions between these universes becomes a legitimate scientific question. Some speculate that paranormal events may be the result of such interactions, such as information transfer between parallel realities or the manifestation of entities or energies from other universes. While this remains highly speculative, the concept of a multiverse offers a framework within which these previously inexplicable phenomena can be considered more seriously.

The investigation into the multiverse remains a challenge for modern physics and cosmology. However, the conceptual framework it offers provides powerful tools for exploring the fundamental nature of reality and the possibility of other forms of existence. As our understanding of quantum mechanics, inflationary cosmology, and string theory deepens, so too will our ability to explore the potential existence and properties of a multiverse, which in turn could illuminate the relationship between our universe and other potential realms of existence, bringing us closer to answering some of humanity's most fundamental questions about the nature of reality itself. The pursuit of such understanding, despite the inherent challenges and speculative nature of the endeavor, is a testament to human curiosity and a driving force behind scientific progress. The continued exploration of these ideas may one day fundamentally reshape our understanding of the universe and our place within it.

The existence of a multiverse, if proven, would profoundly reshape our understanding of reality, dismantling the anthropocentric worldview that places humanity at the center of existence. For millennia, humanity has viewed the universe through a lens of self-importance, assuming our planet and our species held a unique, perhaps even divinely ordained, position in the cosmos. The Copernican Revolution challenged this geocentric view, demoting Earth from the center of the solar system. The discovery of countless galaxies and the vastness of the observable universe further diminished our perceived importance, revealing our planet as a tiny speck in an unimaginably large cosmos. However, even with this expanded cosmic perspective, the notion of a single, unique universe remains deeply ingrained in our thinking. The multiverse hypothesis throws this notion into question, suggesting that our universe may be just one among countless others, each with its own unique properties and perhaps even its own forms of life.

This shift in perspective is not merely a scientific curiosity; it is a profound philosophical upheaval. If our universe is but one of many, the very foundations of our understanding of existence are shaken. The implications extend to our understanding of time, space, causality, and even consciousness itself. The anthropic principle, which suggests that the universe's properties are finely tuned for the emergence of life, gains a new dimension within a multiverse context. In a single universe, this fine-tuning might seem miraculous or divinely orchestrated. However, in a multiverse, the anthropic principle simply reflects the statistical probability that within an infinite or near-infinite number of universes, some will inevitably possess the necessary conditions for life to arise. Our existence, then, is not a unique event, but rather a statistically probable outcome within a vast cosmic lottery.

This understanding leads to a humbling realization of our place in the cosmos. We are not the central players in a cosmic drama, but rather inhabitants of one particular universe, perhaps an unremarkable one among many. This does not diminish the significance of human existence, but rather reframes it within a broader, more expansive context. It invites us to contemplate the potential diversity of life and consciousness across the multiverse, challenging us to expand our imaginations beyond the limitations of our own experience.

The implications of a multiverse for the nature of reality extend beyond our understanding of our place within the cosmos. It raises fundamental questions about the nature of physical laws and constants. In our universe, certain constants, such as the gravitational constant and the speed of light, are fixed. However, in a multiverse, these constants might vary from universe to universe, leading to radically different physical realities. Universes might exist where gravity is much weaker or stronger than in ours, where the speed of light is different, or where the fundamental forces of nature interact in entirely novel ways. Such variations could lead to universes that are fundamentally incompatible with life as we know it, or perhaps to universes where entirely new forms of matter and energy exist. The possibilities are endless, bounded only by the limits of our imagination and our current understanding of physics.

The philosophical implications of a multiverse are equally profound. The concept of multiple universes challenges the very definition of reality. Is reality singular or plural? Are there different levels of reality, with our universe being just one level among many? These questions have occupied philosophers and theologians for centuries, and the multiverse hypothesis offers a new lens through which to examine them. The multiverse potentially challenges traditional notions of causality, as events in one universe might influence events in another, blurring the lines between cause and effect. The very fabric of spacetime, as we understand it, might be fundamentally different in other universes.

Moreover, the existence of a multiverse raises questions about the nature of consciousness. If countless universes exist, each potentially harboring forms of life and consciousness, what does this mean for the uniqueness and value of human consciousness? Is human consciousness a rare and precious phenomenon, or a common occurrence throughout the multiverse? The answers to these questions remain elusive, but the questions themselves highlight the profound implications of the multiverse hypothesis. The very act of questioning the nature of reality, of seeking to understand our place within the vast expanse of the cosmos, is itself a testament to the unique power of human consciousness.

The challenge of verifying the existence of a multiverse is considerable. Currently, we lack the technological capabilities to directly observe or interact with other universes. However, indirect evidence might be found through anomalies in the cosmic microwave background radiation, unexpected fluctuations in the gravitational field, or through theoretical breakthroughs in our understanding of fundamental physics. String theory, for instance, suggests the possibility of a vast "landscape" of universes, each with its own unique set of physical laws and constants. Further developments in string theory and other theoretical frameworks might provide more concrete predictions that could be tested through observation.

Furthermore, the multiverse concept doesn't just have implications for theoretical physics but also for our understanding of the paranormal. If countless universes exist, some with radically different physical laws, the possibility of interactions between these universes becomes a legitimate scientific question. Phenomena currently attributed to the paranormal, such as déjà vu, precognitive experiences, or even purported encounters with extraterrestrial beings, might find a novel explanation within the framework of a multiverse. While this remains highly speculative, it underscores the potential for a multiverse to fundamentally reshape our scientific understanding of reality, bringing formerly unexplained phenomena into the realm of scientific inquiry.

The quest to understand the multiverse is an ongoing journey, pushing the boundaries of human knowledge and challenging our most fundamental assumptions about the nature of reality. The sheer scale of the multiverse hypothesis is daunting, yet the potential rewards of understanding it are immense. It offers a unique opportunity to reassess our place in the cosmos, to re-evaluate our understanding of the physical laws governing the universe, and to perhaps even gain a deeper understanding of the nature of consciousness itself. This exploration is a testament to human curiosity and our relentless pursuit of knowledge, a journey that may ultimately lead to a revolutionary transformation in our understanding of existence. The implications are far-reaching, encompassing not only the physical sciences but also philosophy, theology, and our very understanding of what it means to be human in a potentially infinite universe. The journey to explore the multiverse is a journey into the heart of reality itself.

The sheer scale and strangeness of the multiverse concept naturally lead to the question: can we interact with these other universes? The answer, at present, is a resounding "we don't know," tempered by a cautious "perhaps." Our current understanding of physics doesn't offer a definitive yes or no, but it does provide some intriguing theoretical avenues of exploration. These avenues, however, are firmly rooted in speculation, requiring significant leaps of faith and a willingness to embrace concepts that push the boundaries of our established scientific frameworks.

One of the most popular, and arguably most scientifically plausible, avenues for potential interaction lies in the realm of wormholes. Predicted by Einstein's theory of general relativity, wormholes are hypothetical tunnels through spacetime, connecting two distant points in our universe or even two separate universes. Imagine spacetime as a fabric, and a wormhole as a shortcut, a fold in that fabric that allows for instantaneous travel between vastly separated locations. While theoretically possible, the practical challenges of creating and traversing a wormhole are immense. They require exotic matter with negative mass-energy density, a substance we have yet to discover, to keep the wormhole open and prevent it from collapsing. Furthermore, the immense gravitational forces at play within a wormhole could easily crush anything attempting to traverse it.

Beyond the purely theoretical challenges, the question of navigation through a wormhole remains unanswered. How would one steer a spaceship through such a tunnel? What guarantees that the other end of the wormhole leads to a hospitable location, rather than a black hole or some other inhospitable environment? The very act of navigating a wormhole may represent a technological and engineering feat surpassing anything currently imaginable. Despite these considerable hurdles, the concept of wormholes continues to fuel the imagination of scientists and science fiction writers alike, serving as a potent symbol of the potential for inter-universal travel.

Another fascinating, albeit more speculative, avenue of interaction involves quantum entanglement. This peculiar phenomenon, firmly established through experimental verification, involves two or more quantum particles becoming linked in such a way that they share the same fate, regardless of the distance separating them. Measuring the state of one entangled particle instantaneously reveals the state of the other, even if they are light-years apart. This seemingly instantaneous connection has led some physicists to speculate that quantum entanglement might provide a pathway for communication, or even interaction, between universes.

The idea is that if our universe is entangled with another, perhaps at a fundamental quantum level, then the shared fate of entangled particles could translate to some form of cross-universal influence. This is pure speculation, of course. We don't currently have any evidence suggesting that quantum entanglement extends beyond our universe, and the mechanics of such an interaction remain completely mysterious. However, the strangeness of quantum entanglement, defying classical notions of locality and causality, hints at a deeper reality that might allow for more exotic forms of interaction than we currently comprehend. The implications are staggering, hinting at a potential connection between seemingly disparate universes, potentially bridging the gaps between seemingly separate realities.

The possibility of observing other universes is perhaps even more elusive than interacting with them. One potential method lies in the detection of gravitational waves emanating from other universes. These waves, ripples in spacetime caused by cataclysmic events, could potentially travel between universes, carrying information about their physical properties and inhabitants. However, detecting these faint signals from other universes would require extraordinarily sensitive detectors and a deep understanding of the physics governing inter-universal gravitational wave propagation – an area currently unexplored.

Another intriguing possibility involves searching for evidence of "bubble collisions" within the cosmic microwave background radiation (CMBR), the afterglow of the Big Bang. If our universe collided with another universe in the distant past, it might leave behind detectable imprints on the CMBR. These imprints might manifest as unusual patterns or anomalies in the CMB's temperature distribution, providing indirect evidence of an interaction between universes. The identification and interpretation of such signals, however, requires extremely precise measurements and sophisticated data analysis techniques, and we are yet to find definitive evidence in this area.

Furthermore, the study of variations in fundamental constants across the multiverse could potentially offer clues about the existence and nature of other universes. If the fundamental constants, such as the speed of light or the gravitational constant, are different in other universes, this could leave detectable imprints on the physical structure and evolution of our own universe. Detecting these subtle variations is a significant challenge, and we're only beginning to explore such possibilities through increasingly precise measurements of fundamental constants.

The search for interaction or observation of other universes is thus a multi-pronged endeavor, encompassing several theoretical and observational approaches. While the challenges are immense, and success is far from guaranteed, the potential rewards are equally immense. The very act of searching for evidence of other universes pushes the boundaries of our knowledge and forces us to confront the limits of our current scientific understanding.

It's crucial to maintain a healthy dose of skepticism, however. Many of the concepts discussed here are highly speculative, and the evidence for the existence of other universes remains tentative at best. Moreover, even if we were to discover evidence of interaction or observation, it wouldn't necessarily provide a complete understanding of the multiverse. The vastness and complexity of the multiverse might be so great that our current scientific frameworks are simply inadequate to fully grasp its nature.

The potential for interaction with other universes, whether through wormholes, quantum entanglement, or other yet-undiscovered mechanisms, profoundly impacts our understanding of reality. It compels us to question our anthropocentric biases, to recognize the potential for vastly different forms of life and consciousness existing beyond our own limited perception. The exploration of the multiverse is a journey into the unknown, a challenge to our imagination and a testament to the enduring human quest for knowledge. While the answers remain elusive, the questions themselves are a powerful stimulus for pushing the frontiers of scientific understanding, leading us towards a profound re-evaluation of our place within the cosmos.

The pursuit of inter-universal interaction and observation necessitates a multidisciplinary approach, drawing upon the expertise of physicists, cosmologists, mathematicians, and even philosophers. It's a collaborative endeavor, requiring the synthesis of different perspectives and the integration of various scientific fields. The challenge is not just scientific but also philosophical, as our understanding of the multiverse compels us to re-evaluate our fundamental assumptions about the nature of reality, causality, and consciousness. The exploration of the multiverse is, therefore, a testament to the power of human curiosity and the unending quest to unravel the mysteries of the universe, and perhaps, the universes beyond. The journey itself, fraught with uncertainty and speculation, is perhaps more significant than the destination, pushing the limits of our understanding and shaping our perception of reality itself. This continuous exploration, driven by both theoretical advancements and novel observational techniques, may eventually reveal insights far beyond our current comprehension, potentially revolutionizing not only our understanding of the cosmos but also our very conception of reality. The quest continues, a relentless pursuit of knowledge that may eventually lead us to a profound understanding of the interconnectedness of realities, far surpassing our current imagination.

The exploration of the multiverse naturally leads us to consider a fascinating and controversial area: the potential connection between the multiverse and paranormal phenomena. For centuries, anecdotal accounts of ghosts, telepathy, precognition, and other unexplained events have challenged our understanding of reality. Could these phenomena, often dismissed as pseudoscience or misinterpretations, find a more plausible explanation within the framework of a multiverse?

One compelling line of inquiry involves the concept of "bleed-through" between universes. If our universe is not isolated, but rather exists alongside countless others, the possibility arises that these universes might interact in subtle, unpredictable ways. Imagine the multiverse as a vast ocean of realities, with occasional ripples or waves causing interactions between neighboring universes. These interactions might manifest as unexplained occurrences in our own universe—paranormal events that defy our current laws of physics.

The nature of these interactions is, of course, speculative. One possibility is that variations in fundamental constants between universes could create localized distortions in our reality. For example, a region of spacetime with slightly different gravitational constants in a neighboring universe might create anomalies in our own, manifesting as strange gravitational effects or even the apparent appearance of objects defying gravity. Similarly, variations in the electromagnetic constants could lead to unusual electromagnetic fields, potentially explaining some reports of anomalous energy or electromagnetic disturbances associated with paranormal events.

Another intriguing possibility involves the leakage of information or energy between universes. Imagine a universe with significantly different physical laws, where certain forms of energy or information are more easily transferable than in our universe. This "leakage" might manifest as unexplained energy bursts, anomalous electromagnetic phenomena, or even the transfer of information across universes, potentially explaining events like telepathy or precognition. Of course, the mechanisms underlying such information transfer remain entirely hypothetical, requiring a deeper understanding of quantum entanglement and inter-universal communication.

The concept of parallel realities also offers a potential framework for understanding certain paranormal events. Suppose our universe is just one in an infinite number of parallel realities, each with its own slightly different history and timeline. In some of these realities, events may occur that are essentially "parallel" to those in our own universe, but with subtle differences. These differences could manifest in our reality as anomalies or strange occurrences. For example, a ghost sighting might be explained as a glimpse into a parallel reality where the individual in question is still alive and interacting with our reality's environment in some manner. This requires abandoning the rigid notion of a single, linear timeline and embracing a much more complex, multidimensional reality.

Furthermore, the notion of "residual energy" or information left behind by events in other universes could explain certain paranormal occurrences. Imagine an event in a neighboring universe that leaves behind a subtle imprint in our own universe. This imprint could be a residual energy field, a distortion in spacetime, or even a lingering pattern of information that affects our physical world. This residual energy might manifest as seemingly unexplained phenomena, providing a possible explanation for reports of hauntings or other unexplained occurrences.

The idea of consciousness also deserves consideration in this context. Could consciousness itself exist independently of a physical body, perhaps even bridging the gap between universes? Some theories propose that consciousness is a fundamental aspect of reality, existing across multiple dimensions or even universes. If this is the case, then consciousness might be capable of interacting with other universes or realities, explaining events like near-death experiences, out-of-body experiences, or even spontaneous psychic phenomena. This perspective challenges our conventional understanding of consciousness as solely a product of the brain, suggesting a deeper, more interconnected nature of consciousness that extends beyond our immediate physical reality.

Of course, the connection between the multiverse and paranormal phenomena remains highly speculative. The lack of empirical evidence makes it impossible to definitively prove or disprove such a link. However, by embracing a more open-minded approach, we can explore the possibilities that lie beyond our current scientific frameworks. The study of the multiverse compels us to question our established scientific paradigms, to consider the possibility of realities beyond our current perception.

One significant challenge in exploring this connection is the difficulty in testing these hypotheses. The nature of paranormal phenomena makes them notoriously difficult to replicate or study in a controlled setting. Furthermore, the scale and complexity of the multiverse make direct observation or interaction extremely challenging. While indirect evidence might be found in the form of unusual patterns in the CMB, gravitational anomalies, or unexplained energy events, interpreting these signals remains a significant hurdle. Moreover, the lack of a unified theory of the multiverse hinders our ability to make precise predictions and test the hypotheses linking the multiverse and paranormal events.

Despite these challenges, exploring the possibility of a link between the multiverse and the paranormal provides a compelling framework for potentially explaining some unexplained phenomena. It encourages a more holistic approach to the study of reality, encompassing both established scientific theories and potentially controversial events. This requires a willingness to challenge established paradigms, to entertain unconventional explanations, and to consider the possibility that our understanding of reality is far more limited than we might assume.

The exploration of such a connection, however, necessitates a multidisciplinary approach. Physicists, cosmologists, parapsychologists, and philosophers must collaborate to develop robust theoretical frameworks and experimental approaches. The challenge is to integrate different scientific disciplines, bridging the gap between established scientific knowledge and areas often deemed outside the mainstream. This cross-disciplinary collaboration will be crucial in developing more sophisticated theoretical models and conducting rigorous experimental tests. Moreover, establishing clear criteria for evaluating potential evidence will be vital in preventing bias and ensuring the integrity of the research.

The potential rewards of such an interdisciplinary approach are immense. If we could establish a credible link between the multiverse and paranormal phenomena, it would revolutionize our understanding of reality. It would provide a new framework for understanding consciousness, the nature of reality, and our place within the cosmos. It might even open up new avenues for scientific exploration, potentially leading to technological breakthroughs that currently seem impossible. The journey is fraught with uncertainty, but the potential implications are profound enough to warrant a comprehensive and rigorous investigation. Even if we fail to find conclusive evidence, the pursuit itself will push the boundaries of our scientific understanding and challenge us to think more creatively about the nature of reality. The exploration is, in itself, a testament to humanity's enduring curiosity and our relentless pursuit of understanding the universe in all its enigmatic complexity. This pursuit may ultimately reshape our perception of reality itself, transcending the limitations of our current scientific paradigms and opening up entirely new avenues for scientific discovery and philosophical inquiry. The quest continues, driven by the inexhaustible human desire to comprehend the mysteries of existence, both within and beyond our perceived reality.

The search for evidence supporting the multiverse hypothesis is a daunting task, given the inherent difficulties in observing or interacting with other universes. However, several avenues of inquiry offer tantalizing clues, albeit indirect and often debated. One promising area lies in the realm of cosmology and astronomical observations. Specifically, the cosmic microwave background (CMB) radiation, the afterglow of the Big Bang, has revealed subtle anomalies that some researchers interpret as potential signatures of interactions with other universes. These anomalies, such as cold spots or unusual temperature fluctuations, defy simple explanations within the framework of a single universe model. However, it's crucial to acknowledge that alternative explanations, such as unknown physical processes within our own universe, exist, making definitive conclusions challenging.

The study of the CMB also involves intricate modeling and simulations. These simulations attempt to recreate the universe's evolution from its earliest moments, incorporating various cosmological parameters. In some simulations, scenarios involving multiverse interactions produce CMB patterns that surprisingly resemble observed anomalies. While these simulations offer suggestive evidence, it's vital to remember that the models themselves are subject to assumptions and uncertainties regarding the initial conditions and physical laws governing the universe's evolution. The degree to which these simulations accurately reflect reality is still a matter of ongoing investigation and debate.

Beyond the CMB, certain aspects of large-scale structure formation in the universe might hint at the existence of a multiverse. The distribution of galaxies, galaxy clusters, and voids in the observable universe exhibits patterns that some researchers suggest are inconsistent with a standard cosmological model of a single universe. However, again, interpreting these observations is complex, requiring careful consideration of observational biases, measurement errors, and the inherent limitations of our current understanding of the universe's formation and evolution. These large-scale structure features might be better explained by alternative theories of gravity or dark matter distribution without necessarily invoking a multiverse.

Theoretical predictions also play a crucial role in the search for multiverse evidence. String theory, a leading candidate for a theory of quantum gravity, predicts the existence of a vast landscape of possible universes, each with different physical constants and laws. This landscape of universes naturally lends itself to a multiverse scenario. However, the mathematical complexities of string theory make it difficult to extract testable predictions that definitively confirm or refute its predictions. The absence of experimental verification, despite decades of research, remains a major hurdle in establishing string theory as a robust model for the multiverse.

Another theoretical approach involves eternal inflation, a cosmological model suggesting that inflation, the period of rapid expansion in the early universe, may never truly end. In this scenario, pockets of spacetime continuously inflate, creating a vast network of universes, each with its own unique physical characteristics. This model offers a plausible mechanism for the multiverse's creation but lacks direct observational support. Detecting or observing these constantly inflating regions remains well beyond our current technological capabilities. Furthermore, testing the predictions of eternal inflation against observations requires sophisticated cosmological simulations and theoretical modeling, the results of which remain highly uncertain.

The lack of direct observational evidence for a multiverse doesn't automatically invalidate the hypothesis. The sheer scale and nature of the multiverse, encompassing realms potentially beyond our current understanding of space and time, make direct observation extremely difficult, if not impossible. Moreover, the limitations of our current observational techniques and theoretical frameworks might restrict our ability to detect the subtle signals of a multiverse's existence. The possibility of universes beyond the scope of our detection should be acknowledged.

Some physicists have argued that the fine-tuning of the universe's physical constants – the values of parameters that determine the fundamental forces and properties of the universe – provides indirect evidence for a multiverse. The argument goes that if these constants differed even slightly, our universe would be radically different, perhaps unable to support life. In a multiverse, one might expect a vast range of universes with varying constants, with our universe being merely one of the rare instances where conditions are suitable for life's emergence. While this "anthropic principle" offers a plausible explanation for the fine-tuning, it also remains highly controversial. It's difficult to test or falsify, making it a philosophical argument rather than a directly verifiable scientific hypothesis.

It is also essential to consider the alternative explanations that challenge the multiverse hypothesis. Many of the observations sometimes interpreted as evidence for a multiverse can be explained within the framework of existing cosmological models, albeit with modifications or extensions. For instance, the CMB anomalies can be attributed to unknown physical processes or observational biases, as mentioned earlier. Similarly, the large-scale structure of the universe can be modeled effectively without invoking a multiverse. It is crucial for scientific rigor to explore and exhaust alternative interpretations before invoking a more extraordinary hypothesis.

The exploration of the multiverse requires a multidisciplinary approach, drawing insights from cosmology, particle physics, string theory, and other relevant fields. Researchers must collaborate to develop more sophisticated models, devise innovative experimental techniques, and critically evaluate existing data. The current lack of definitive evidence doesn't diminish the importance of continuing the pursuit. The search for the multiverse is a testament to human curiosity and the relentless drive to unravel the deepest mysteries of the cosmos.

Furthermore, the philosophical implications of the multiverse hypothesis are profound. If multiple universes exist, each with its own set of physical laws and perhaps even different forms of life, our understanding of our place in the cosmos undergoes a fundamental shift. The seemingly unique and privileged position of humanity in the universe might be diluted, leading to new perspectives on existence, consciousness, and the meaning of life itself. This philosophical exploration is an integral aspect of studying the multiverse.

The exploration of the multiverse should not be viewed as mere speculation but as a crucial scientific pursuit, even if currently limited by technological and theoretical constraints. The potential scientific and philosophical breakthroughs associated with understanding the nature of the multiverse are significant enough to warrant dedicated, continued research. The path towards understanding the multiverse is filled with challenges, but the potential rewards are immense. The journey may be long, and the answers might remain elusive for many years, but the quest itself is a testament to humanity's enduring fascination with the universe and our place within it. It is a testament to the ongoing scientific endeavor that relentlessly pushes the boundaries of knowledge, even when confronting the most enigmatic and challenging questions about the nature of reality. This pursuit will undoubtedly reshape our perception of reality itself, potentially revolutionizing our understanding of the universe and our position within the cosmos.

5

Ghosts and Otherworldly Encounters

The intriguing possibility of a multiverse, with its implications for the nature of reality, naturally leads us to consider phenomena that seem to defy conventional scientific explanation. One such area is the investigation of ghostly phenomena, often relegated to the realm of pseudoscience but worthy of a rigorous, scientific inquiry, however challenging. While the existence of ghosts remains unproven, the pursuit of empirical evidence necessitates a careful examination of the methods used and the conclusions drawn.

Many attempts to investigate ghostly phenomena have centered on the detection of anomalous electromagnetic fields (EMFs). The rationale behind this approach stems from the idea that ghosts, if they exist, might interact with their environment through subtle energy fluctuations, potentially detectable as variations in EMF readings. Various instruments, such as EMF meters and magnetometers, have been employed in haunted locations to search for such anomalies. However, the results of these investigations have been, at best, inconclusive and often plagued by methodological shortcomings.

One major challenge is the difficulty in controlling for extraneous variables. EMFs are ubiquitous in our environment, generated by numerous sources, including electrical wiring, appliances, and even natural phenomena. Differentiating a genuine paranormal signal from background noise, or artifacts from the measuring equipment itself, proves exceedingly difficult. Without meticulous calibration, shielding, and control experiments, any detected EMF anomalies could easily be attributed to mundane sources. Studies often fail to adequately address these crucial control parameters, leading to weak and unconvincing evidence.

Furthermore, the interpretation of EMF readings requires careful consideration. Even if a statistically significant anomaly is detected, it doesn't automatically imply a paranormal cause. There are numerous potential explanations, including faulty instrumentation, electrical interference, or even subtle geological phenomena affecting the local EMF. The absence of a clear, consistent correlation between EMF anomalies and reported ghostly activity further weakens the case for a paranormal link. Many studies have reported EMF spikes in purportedly haunted locations but fail to establish a causal relationship with the reported events, rendering the findings inconclusive and open to alternative interpretations.

Beyond EMF studies, other approaches to investigating ghostly phenomena have employed various methodologies, with varying degrees of scientific rigor. Infrared thermography, for instance, has been used in an attempt to detect temperature fluctuations associated with supposed ghostly presences. However, similar challenges to EMF studies arise, namely, the difficulty in controlling for extraneous sources of temperature variation and distinguishing a subtle anomaly from natural background fluctuations. Without robust controls and careful analysis, any observed temperature anomalies remain ambiguous and unable to provide definitive evidence for paranormal activity.

Psychological factors also play a critical role in many alleged ghostly encounters. The suggestibility of witnesses and the power of suggestion can significantly impact observations. The confirmation bias, the tendency to seek out and interpret information that supports pre-existing beliefs, can lead to misinterpretations of ambiguous events. Furthermore, the social context of ghost hunting investigations can amplify the effects of these biases. The expectations and beliefs of the investigators themselves can influence their perceptions and interpretations of the data, potentially leading to biased observations and conclusions. Objective, double-blind studies, where investigators are unaware of the experimental conditions, are crucial to mitigate these subjective influences.

Furthermore, the phenomenon of pareidolia, the tendency to perceive meaningful patterns in random stimuli, is often relevant in ghost investigations. The human brain is remarkably adept at finding patterns, even when they are not actually present. This ability, while generally beneficial, can lead to misinterpretations of ambiguous visual or auditory data, particularly in dimly lit or acoustically challenging environments. A creaking floorboard can be misinterpreted as a ghostly footstep, and a shadow can take on an otherworldly form fueled by pre-existing beliefs and expectations. These perceptual biases need to be carefully considered when analyzing evidence of ghostly phenomena.

The use of audio recordings and video footage also presents challenges. Audio recordings can be easily manipulated or misinterpreted. Background noises, poorly calibrated equipment, or even simple misunderstandings can lead to spurious conclusions. Similarly, video footage can be susceptible to visual artifacts, poor lighting conditions, or camera malfunctions, leading to the misinterpretation of ordinary occurrences as paranormal events. Furthermore, digital manipulation of such recordings is now easily performed and can be used to fabricate or distort evidence, rendering such recordings unreliable unless verifiable and independently confirmed.

Even more sophisticated investigations employing sophisticated technologies still face significant methodological challenges. For example, the use of advanced imaging techniques or other sensitive instruments can inadvertently introduce noise or artifacts that are difficult to distinguish from genuine paranormal signals. The interpretation of such data often relies on complex algorithms and models, which can themselves introduce biases or errors. Any investigation claiming to provide evidence of ghostly phenomena must rigorously demonstrate the reliability and validity of its methodology and address the potential for these sources of error.

In conclusion, while the possibility of ghostly phenomena cannot be definitively ruled out, the current scientific evidence remains inconclusive. While some studies have reported anomalies in various physical parameters, the methodological limitations and potential alternative explanations render these findings insufficient to establish a firm link to the paranormal. Rigorous scientific investigations of ghostly phenomena must address the critical issues of experimental control, observer bias, and the potential for spurious interpretations of data. Until these challenges are overcome, and the methodology meets the highest standards of scientific rigor, the existence of ghosts remains an intriguing yet unproven hypothesis. The continued pursuit of rigorous scientific inquiry, however, remains vital in further exploring these enigmatic phenomena, even if the path to definitive answers may prove long and challenging. Only through such meticulous and unbiased investigations can we hope to either confirm or definitively refute the existence of ghostly phenomena.

Having considered the methodological limitations of previous attempts to scientifically investigate ghostly phenomena, we now turn our attention to a more theoretical, yet potentially fruitful avenue of inquiry: the concept of density variations as a possible explanation for ghostly encounters. This approach departs from the largely empirical investigations discussed previously, instead focusing on a theoretical framework grounded in physics and cosmology. The fundamental idea is that different regions of spacetime might exhibit varying densities, leading to the creation of what could be described as "density realms" – regions with physical properties sufficiently different from our own to appear as separate planes of existence.

This concept draws upon established physics, particularly Einstein's theory of general relativity, which describes gravity as a curvature of spacetime caused by mass and energy density. Regions of high density warp spacetime significantly more than regions of low density. Extrapolating from this, we might envision regions of spacetime with extremely low or extremely high density, potentially so different from our own as to create pockets of existence virtually invisible to us under normal circumstances. These regions could possess different physical laws, different temporal flows, and possibly even different fundamental constants.

One intriguing possibility is that these density realms could be partially or intermittently accessible to our own, creating the illusion of ghostly apparitions or otherworldly encounters. A localized anomaly, a fluctuation in density, could temporarily bridge the gap between our reality and another, allowing for glimpses of – or interactions with – entities inhabiting this different density realm. This concept isn't necessarily proposing a complete alteration of physics, but rather exploring how variations within the established framework could give rise to phenomena currently unexplained. Imagine, for example, a ripple in spacetime, a momentary fluctuation in density that creates a "window" into another realm, allowing a temporary manifestation of a being from that realm into our own. This manifestation might appear as a ghostly figure, exhibiting properties inconsistent with our understanding of the physical world – such as passing through walls, appearing and disappearing abruptly, or communicating in ways defying typical physical laws.

The concept of density variations finds a parallel in some cosmological models describing the early universe. The universe began in an extremely dense state, and its subsequent expansion and cooling have resulted in the highly structured and less dense cosmos we observe today. It's not inconceivable that pockets of higher density might persist even today, representing remnants of the early universe's denser state. These pockets, embedded within the overall fabric of spacetime, could be relatively stable and potentially harbor life forms adapted to these extreme conditions. If so, these life forms might only become briefly perceivable to us under specific circumstances, such as during shifts or fluctuations in the local density, giving rise to the phenomena often associated with ghosts.

It is crucial to emphasize that this is a highly speculative concept, requiring significant further theoretical and experimental exploration. However, it does offer a potentially testable hypothesis. For instance, if ghostly apparitions are indeed linked to density variations, we might expect to observe correlations between reported paranormal activity and fluctuations in local gravitational fields, electromagnetic fields, or other physical parameters sensitive to changes in density. Precise measurements of gravity and subtle shifts in local spacetime curvature at locations associated with reported hauntings could help test this hypothesis. High-precision gravimeters, highly sensitive magnetometers, and other sophisticated instrumentation could be utilized to gather data at such locations, searching for correlations with reported events.

The challenge lies in designing experiments that effectively control for extraneous variables, such as geological factors that can influence local gravity or electromagnetic fields. Furthermore, the very subtle nature of the density variations we might be searching for implies that the sensitivity of the instruments must be exceptionally high, necessitating substantial technological advancements. Nevertheless, the potential payoff—an entirely new understanding of reality, its composition, and the possibility of other forms of life—is well worth the effort.

The interpretation of any such data would also need to be carefully approached, acknowledging the potential for spurious correlations and the influence of observational biases. Double-blind studies, where researchers are unaware of the purportedly haunted location or the timing of reported events, could significantly mitigate these issues. Advanced statistical techniques, designed to control for multiple variables and identify meaningful patterns amidst noise, would be crucial in interpreting the findings.

This framework offers a different perspective on the problem of ghosts, moving away from the traditionally employed methodologies of EMF measurements and anecdotal accounts. It encourages a shift towards a more physics-based investigation, seeking to understand the underlying physical processes that might give rise to these paranormal occurrences. This doesn't negate the possibility that some reported ghostly phenomena might have purely psychological explanations, but it opens a new and scientifically rigorous avenue for exploration. Instead of dismissing these reports outright, we can instead reframe them within the context of our expanding understanding of the universe and the potential for a far more complex and dynamic reality than previously imagined.

Furthermore, this approach encourages a deeper understanding of the very fabric of reality. It pushes us to consider the possibility that our perception of the universe, with its seemingly fixed laws and limitations, might be far more limited than we currently believe. The existence of density variations or fluctuations in spacetime could profoundly alter our understanding of the universe's structure and potentially the nature of consciousness itself. It opens up the possibility that consciousness might not be uniquely tied to the specific conditions of our own density realm and that other forms of consciousness might exist within different density realms, manifesting in ways that appear inexplicable to us.

The implications of this concept extend far beyond the realm of ghost hunting. If different density realms exist, and if they are even intermittently accessible, it could revolutionize our understanding of cosmology and astrophysics. The search for extraterrestrial life, for instance, might need to expand beyond our current search strategies, which are often limited by our assumptions about the physical conditions necessary for life. If other realms exist with different physical laws and constants, life could exist in ways unimaginable to us, potentially hidden in plain sight yet undetectable with our current methods. The detection of such density variations, therefore, could potentially hold the key not only to understanding the nature of ghostly phenomena but also to unlocking new frontiers in our exploration of the universe and our place within it.

The concept of density variations also challenges our very definition of reality. What we perceive as solid matter is ultimately a manifestation of concentrated energy and matter interacting within a specific density range. If different density realms exist, they could contain forms of matter and energy unknown to us, interacting under drastically different conditions. Our current scientific models and instruments might be incapable of detecting or interacting with such forms of matter and energy. This opens the intriguing possibility that what we consider "empty space" might actually be teeming with other realities, other forms of matter and energy, existing in different density realms, beyond our current perception and understanding.

The exploration of density variations as a potential explanation for ghostly encounters, therefore, represents a significant challenge but also a potentially profound opportunity. It necessitates a multidisciplinary approach, bringing together physicists, cosmologists, and parapsychologists to develop new theoretical models, design sophisticated experimental techniques, and critically evaluate the data obtained. The path forward is undoubtedly challenging, requiring significant technological advancements and a willingness to explore beyond the boundaries of conventional scientific thinking. However, the potential rewards—a richer, more complete understanding of reality and our place within it—make this a worthy pursuit. The journey may be long and arduous, but the potential destination is a paradigm shift in our understanding of the universe, a shift that could encompass not only ghostly phenomena but also the very essence of reality itself.

Let us now delve into specific documented cases of ghostly encounters, examining them through the lens of density variations. While acknowledging the inherent limitations in rigorously verifying anecdotal evidence, these case studies offer valuable opportunities to explore the plausibility of our proposed hypothesis. The goal is not to definitively prove the existence of density realms, but to assess whether these reported experiences might be consistent with the theoretical framework we've established.

One frequently cited case involves the alleged hauntings at Borley Rectory in England. Numerous witnesses reported various paranormal phenomena over several decades, including ghostly apparitions, unexplained noises, and objects moving on their own. While many dismissed these events as hoaxes or misinterpretations, some accounts suggest a pattern of activity linked to specific locations within the rectory, suggesting a possible spatial correlation with potential density fluctuations. If our hypothesis is correct, we might expect a concentration of reported events in areas with subtle variations in gravitational or electromagnetic fields – areas where the "fabric" of spacetime itself might be experiencing localized stress or distortion, facilitating intermittent access to another density realm. Detailed historical accounts of Borley Rectory's construction and geological features would be necessary to correlate these reported events with any potentially measurable physical anomalies. We need to investigate the possibility that unseen geological formations beneath the rectory could cause localized density variations, potentially acting as foci for paranormal activity.

Another intriguing case is that of the Amityville Horror. While the validity of this case remains highly debated, the reported phenomena – disembodied voices, unexplained temperature changes, and the sensation of unseen presences – could potentially be interpreted within our density variation framework. It's crucial to carefully examine the available evidence, differentiating between documented events and unsubstantiated claims. However, if even a subset of the reported events can be reliably verified, then analyzing the spatial and temporal patterns of these occurrences in relation to any measurable physical anomalies could provide valuable insights. This might involve revisiting the Amityville house and using high-precision instruments to detect minute fluctuations in gravitational fields, electromagnetic fields, or even subtle variations in spacetime curvature.

The numerous reports of poltergeist activity, characterized by unexplained object movement and other disruptive phenomena, also merit attention. These cases, often associated with specific individuals or locations, might suggest localized density fluctuations impacting the physical world. We must acknowledge the possibility of psychological factors playing a role in these events, but it's also possible that some instances of poltergeist activity could be connected to temporary breaches in our own density realm and another. The energy involved in such a breach, even if minuscule, could potentially manifest as the unexplained movement of objects. This would require a careful examination of poltergeist case studies, specifically focusing on instances where robust documentation exists, separating verifiable events from subjective experiences.

Consider, for instance, the Enfield Poltergeist case, which garnered significant attention in the 1970s. The reported phenomena, including object levitation, disembodied voices, and physical assaults, were investigated by reputable researchers. While skepticism remains, a re-examination of the documented evidence from this case, focusing on spatial and temporal correlations with any measurable physical anomalies, could provide further insight into the plausibility of the density variation hypothesis. This would involve not only revisiting the Enfield house but also carefully analyzing the original documentation to pinpoint the spatial locations of the reported events with the highest precision possible, along with temporal data, to look for possible correlations.

Many haunted locations worldwide feature recurring reports of similar phenomena—unexplained sounds, flickering lights, and feelings of cold spots. While these occurrences might often be attributed to natural causes or psychological factors, some cases present patterns that are difficult to explain conventionally. For example, consistent reports of cold spots in specific areas of a building could be indicative of a localized energy anomaly related to density fluctuations. A cold spot might represent a region where energy is being "drawn away" from our realm into another, creating a localized decrease in temperature. Similarly, unexplained sounds could be attributed to vibrations or energy transfers between different density realms. Systematic mapping of these phenomena in conjunction with detailed physical measurements could reveal underlying patterns, suggesting a possible correlation with density variations.

In analyzing these case studies, it's crucial to avoid confirmation bias. We must remain open to alternative explanations, such as misinterpretations, psychological factors, and even outright hoaxes. The strength of our hypothesis lies not in proving the existence of ghosts, but in offering a scientific framework for understanding reported anomalous occurrences. By applying a multi-disciplinary approach, combining rigorous scientific methods with careful consideration of anecdotal evidence, we can move beyond subjective interpretations and explore the possibility of objective physical phenomena underlying these experiences. This includes rigorous investigation into the psychological state of witnesses, and eliminating all possible mundane explanations.

It is also vital to acknowledge the limitations of current technology. Our ability to detect subtle variations in spacetime density is still in its infancy. However, ongoing advancements in areas like gravitational wave detection and high-precision metrology offer hope for developing more sophisticated instruments capable of detecting these extremely subtle variations. Such instruments could be used to systematically survey purportedly haunted locations, searching for correlations between reported paranormal activity and measurable physical anomalies. This could revolutionize the study of parapsychology, moving it away from mere anecdotal evidence towards a quantitative, physics-based approach.

The systematic application of advanced statistical methods will also be crucial in this endeavor. With the vast number of variables involved, simple correlation analysis is insufficient. Sophisticated algorithms that account for confounding factors and allow us to tease out meaningful patterns from noise are required. The development of these statistical methods is a crucial next step in developing a scientifically sound approach to investigating these phenomena. This includes developing methods to account for potential biases introduced by subjective reporting and psychological influence on the witnesses.

The path forward requires a multi-disciplinary collaboration. Physicists can develop sophisticated models for density variations and design experiments to detect them. Cosmologists can provide insights into the potential origins of such density variations within the context of the universe's evolution. And parapsychologists can contribute with their expertise in collecting and analyzing data related to paranormal occurrences. Such collaboration would create a synergistic approach, potentially leading to a breakthrough in our understanding of reality. The ultimate goal isn't necessarily to prove or disprove the existence of ghosts, but rather to expand our scientific understanding of the universe and the possible interactions between different realms of existence.

The case studies themselves, while inherently anecdotal, provide a valuable starting point for a more rigorous investigation. By analyzing these cases systematically, focusing on correlations with measurable physical variables, and deploying advanced statistical techniques, we might begin to unveil a deeper understanding of the phenomena often attributed to the paranormal. The exploration of density variations as a potential explanatory framework offers a fresh perspective, moving beyond simplistic dismissals and encouraging a more open-minded, scientifically rigorous approach to investigating the enigmatic world of ghostly encounters. This holistic approach offers a chance to potentially merge theoretical physics, cosmology, and parapsychology, potentially revolutionizing our comprehension of reality itself.

Let's expand on the concept of density variations as a potential explanation for ghostly experiences. We've explored how fluctuations in spacetime density, even at a microscopic level, could create temporary "windows" or "tunnels" between different density realms. These realms, according to our hypothesis, aren't necessarily separate universes in the conventional multiverse sense, but rather variations within our own universe, existing in a state of superposition or entanglement with our perceived reality. The difference lies in the density of spacetime itself; a denser region might manifest as a perceived "otherworldly" realm, while a less dense area might be undetectable to our senses.

Imagine spacetime as a fabric, not uniformly smooth but possessing subtle wrinkles and folds. These wrinkles, representing variations in density, could act as conduits for energy and even matter to transit between different density regions. A sudden increase in density in a specific location could cause a temporary disruption in the "fabric" of our reality, allowing energy from a denser realm to momentarily bleed into ours.

This energy, carrying information and patterns from the other realm, might manifest as what we perceive as a ghost—a fleeting imprint of a being or event existing in a different density state.

This doesn't necessarily imply the existence of disembodied spirits in the traditional sense. Instead, the "ghostly" apparition might be a manifestation of residual energy or information from another density realm, a sort of echo or reverberation from a past event or entity. The reported characteristics of ghosts – their transparency, their ability to pass through walls, their sometimes-indistinct forms – could all be explained by this phenomenon. The energy leaking into our realm wouldn't possess the solidity of our physical world; it would be a diluted, less-dense manifestation of a phenomenon originating in a higher-density region.

The intermittent and unpredictable nature of ghostly encounters further supports this hypothesis. The "windows" between density realms, due to their dependence on subtle spacetime fluctuations, wouldn't be permanently open. They would exist only fleetingly, triggered by unknown physical processes. This might explain why some haunted locations exhibit activity only sporadically, or why encounters are often brief and difficult to replicate.

Furthermore, the reported emotional or psychological impact of ghostly encounters might also find a physical explanation within this framework. The transfer of energy between realms could induce electromagnetic disturbances, potentially affecting brain activity and eliciting strong emotional responses in witnesses. The sense of "cold spots" often associated with paranormal activity could be attributed to a localized energy drain, as energy is drawn from our realm into the denser region. The unexplained sounds associated with ghostly experiences could be due to vibrations or energy transfers through the "tunnels" between realms, generating subtle sonic phenomena.

Beyond density fluctuations, we must also consider the possibility of interdimensional interactions. While the concept of interdimensional travel is often relegated to science fiction, our expanding understanding of quantum mechanics and the multiverse suggests that such interactions might be possible, albeit on a scale far smaller than we typically imagine. Minute quantum fluctuations could cause temporary breaches in the dimensional fabric, creating pathways for energy or information to traverse between different dimensions. These breaches might be extremely localized and fleeting, explaining the transient nature of many paranormal encounters.

The interaction between these different realms might not necessarily be one-way. Just as energy from a denser realm might bleed into ours, it's conceivable that energy from our realm could influence the denser realms. This opens the possibility of a two-way interaction, where human thoughts, emotions, and even intentions could influence the phenomena occurring in these other realms. This is a fascinating avenue for exploration, particularly in relation to phenomena such as poltergeist activity, where the activity often seems to be tied to the emotional or psychological state of individuals present.

Applying our density-variation framework, we can reinterpret previously unexplained phenomena. Take, for instance, the reported cases of objects moving spontaneously, levitation, or even physical assault associated with poltergeist activity. These seemingly inexplicable occurrences could be manifestations of energy transfers between different density realms. A sudden surge of energy from a higher-density realm, through a temporary breach, could cause objects to move, levitate, or even inflict physical force. This energy wouldn't be a conscious force, but a raw, uncontrolled manifestation of energy from another state of existence.

Another compelling aspect is the subjective nature of paranormal experiences. What one person perceives as a ghostly apparition, another might dismiss as a trick of the light or a figment of their imagination. This variability doesn't necessarily invalidate the experiences; it highlights the subjective nature of perception. Our senses are tuned to a particular range of physical stimuli. We are not equipped to directly perceive phenomena existing outside this range, but they might still affect us indirectly, manifesting as subtle sensory distortions or subconscious impressions.

Furthermore, the cultural and societal context in which ghostly encounters are reported plays a crucial role. The interpretation and understanding of these events are often shaped by prevalent beliefs and cultural narratives. A flickering light in an old house might be interpreted as a paranormal event in a culture where ghost stories are prevalent, while it might be attributed to faulty wiring in a more scientifically-oriented society. This doesn't diminish the reality of the physical phenomena, but highlights the influence of cultural lenses in interpreting those phenomena.

The development of new technologies and scientific tools is crucial for further investigating these phenomena. Advancements in gravitational wave detection, high-resolution electromagnetic sensors, and quantum computing could provide unprecedented opportunities to detect and analyze the subtle energy fluctuations predicted by our density variation hypothesis. These instruments could be deployed in locations associated with paranormal activity, allowing for the quantitative measurement of previously unmeasurable phenomena. The goal is not to "prove" the existence of ghosts, but to develop a robust scientific framework to understand and explain the reported occurrences.

The investigation of these phenomena demands a rigorous interdisciplinary approach. Physicists need to develop theoretical models to predict the measurable consequences of density variations; cosmologists can provide a broader context within the larger framework of the universe; and parapsychologists can contribute their expertise in collecting and analyzing data related to paranormal experiences. This collaboration promises to not only advance our understanding of these seemingly inexplicable phenomena but also to broaden our knowledge of the universe and our place within it.

The journey towards understanding these phenomena is one of meticulous investigation, open-mindedness, and interdisciplinary collaboration. It's a quest that requires us to challenge conventional thinking, embrace the unknown, and develop innovative methodologies to address age-old mysteries. By adopting a scientific, yet open-minded, approach, we can potentially unlock profound insights into the nature of reality itself, bridging the gap between science and the seemingly paranormal. The ultimate goal is not to prove or disprove the existence of ghosts, but to understand the fundamental physical processes underlying these reported occurrences, expanding our scientific understanding of the cosmos and our place within it. This holistic approach transcends the limitations of simple explanations and fosters a comprehensive understanding of reality, one that embraces both the observable and the yet-to-be-understood.

The human mind, a marvel of complexity, is also prone to misinterpretations and distortions of reality. Our perceptions are not always accurate reflections of the external world; they are filtered through our individual experiences, biases, and expectations. This inherent subjectivity plays a significant role in shaping our beliefs about ghosts and other paranormal phenomena. Consider, for example, the well-documented phenomenon of pareidolia, the tendency to perceive meaningful patterns in random stimuli. A creaking floorboard in an old house might be interpreted as a ghostly presence by someone predisposed to believing in ghosts, while another might simply attribute it to the natural settling of the building. This difference in interpretation stems not from a difference in the sensory input itself, but from a difference in the pre-existing mental framework through which that input is processed.

Memory, too, is a fallible mechanism. Memories are not static recordings of events but rather reconstructive processes, influenced by subsequent experiences and beliefs. A vague sensory impression, such as a fleeting shadow or a whisper of sound, might be embellished and transformed over time into a vivid and detailed ghost story, as the mind attempts to impose meaning and coherence upon ambiguous sensory information. This process is particularly potent when coupled with suggestion. The influence of others, be it through storytelling, shared experiences, or media portrayals, can significantly shape our memories and interpretations of events. A group of friends exploring a supposedly haunted location, for instance, might each experience slight sensory anomalies, but through mutual reinforcement, these experiences can be transformed into a compelling collective narrative of a paranormal encounter. The power of suggestion is amplified when the individuals involved share a common belief in the paranormal.

Cultural factors further influence our interpretation of ghostly encounters. In societies where ghost stories are prevalent, and beliefs in the afterlife are deeply ingrained, individuals might be more likely to attribute ambiguous sensory experiences to paranormal causes. The cultural narratives surrounding ghosts shape not only the interpretation of sensory data but also the very expectations individuals have when encountering unusual events. A culture rich in folklore concerning spectral beings might condition individuals to expect or even anticipate such occurrences, increasing the likelihood of misinterpreting ordinary phenomena as paranormal events. This cultural influence underscores the importance of considering the societal and historical context when evaluating accounts of paranormal experiences.

The psychological phenomenon of confirmation bias further complicates the matter. Once an individual has formed a belief in ghosts, they tend to selectively focus on information that supports that belief, while ignoring or dismissing evidence that contradicts it. This selective attention can reinforce and solidify pre-existing beliefs, making them resistant to change. A person convinced of a particular location's haunted status might interpret every unexplained noise or odd sensation as proof of ghostly activity, while discounting rational alternative explanations. This confirmation bias can create a self-reinforcing cycle, where belief strengthens through selective perception and interpretation.

Beyond confirmation bias, the human mind is adept at finding patterns and connections even when none exist. This predisposition, known as apophenia, can lead to the misinterpretation of random coincidences as meaningful events. A series of seemingly unrelated occurrences, such as flickering lights, unexplained noises, and a sudden drop in temperature, might be interpreted by a believer in the paranormal as a coordinated manifestation of ghostly activity, even though these events may be completely unrelated. The human mind's capacity for pattern recognition, a crucial survival skill, can be misleading when applied to the ambiguous realm of paranormal investigations.

Furthermore, the role of anxiety and fear should not be overlooked. Fear can profoundly distort perception and memory, making individuals more susceptible to misinterpreting sensory information. In a dimly lit, supposedly haunted location, a sudden noise might be amplified by fear, leading to a heightened sense of unease and the interpretation of the noise as a ghostly presence. This emotional element of paranormal experiences reinforces the subjective nature of these events, highlighting the limitations of relying solely on subjective accounts as evidence for paranormal activity.

Moreover, the power of suggestion should not be underestimated. Through deliberate or unintentional cues, individuals can be influenced to perceive or remember events differently than they actually occurred. This influence can be particularly potent in group settings, where shared expectations and social pressures can lead to a convergence of perceptions and interpretations. Hypnosis and other suggestibility-enhancing techniques can exacerbate this effect, further complicating the evaluation of paranormal claims based on eyewitness accounts.

Cognitive biases and the limits of human perception can significantly influence the interpretation of paranormal encounters. The human brain is not a perfectly accurate instrument of perception, but rather a complex information-processing system prone to errors and distortions. Understanding these limitations is crucial for a balanced and nuanced perspective on ghostly beliefs.

Beyond the individual psychological factors, the social context surrounding paranormal beliefs should also be considered. The social transmission of beliefs, through stories, narratives, and cultural traditions, plays a vital role in shaping perceptions and experiences. The shared belief in ghosts within a community can reinforce these beliefs through social validation and consensus. This collective reinforcement can be self-perpetuating, creating a cultural environment where such beliefs are normalized and accepted.

Investigating the psychology of ghostly beliefs requires an interdisciplinary approach, drawing upon the fields of psychology, cognitive science, and sociology. A multi-faceted investigation can offer a more holistic understanding of why individuals believe in ghosts and the influence of psychological and social factors in shaping these beliefs. By disentangling the complex interplay of perception, memory, suggestion, and cultural influences, we can begin to develop a more comprehensive perspective on the psychological underpinnings of ghostly beliefs, enabling a more rigorous and evidence-based approach to exploring the phenomena associated with these beliefs. This scientific approach, while acknowledging the power of suggestion and cultural context, allows for a clear-eyed exploration of the potential physical phenomena that might underlie the reported experiences, bridging the gap between subjective accounts and objective scientific investigation. The ultimate goal, as always, remains a rigorous and unbiased investigation into the nature of reality.

6

Extraterrestrial Life and Other Realities

The search for extraterrestrial life represents one of humanity's most ambitious and enduring scientific quests. For centuries, the question of whether we are alone in the universe has captivated our imaginations, fueling countless science fiction stories and inspiring generations of scientists and researchers. The modern scientific approach to this question largely centers around the Search for Extraterrestrial Intelligence (SETI), a multifaceted endeavor employing a range of techniques to detect signs of intelligent life beyond Earth.

SETI's primary focus is the detection of artificial radio signals, a technological signature that would strongly suggest the presence of a technologically advanced civilization. The rationale behind this approach is straightforward: if another civilization has developed radio technology, it's plausible that they might be transmitting signals, either intentionally or unintentionally, that we could detect with our increasingly sophisticated radio telescopes. These telescopes, like the Allen

Telescope Array and the Very Large Array, are capable of scanning vast swathes of the sky, searching for anomalous signals that stand out from the natural background noise of the universe. The challenge, however, lies in distinguishing genuine extraterrestrial signals from terrestrial interference, a task made significantly more complex by the sheer volume of radio waves generated by human activities.

The search for extraterrestrial radio signals requires immense computational power to process the massive amounts of data collected by radio telescopes. Sophisticated algorithms are employed to identify patterns and anomalies that might indicate an artificial origin. The sheer scale of the search necessitates collaboration between multiple research institutions and scientists worldwide, pooling resources and expertise to maximize the chances of success. Despite decades of effort, no conclusive evidence of extraterrestrial radio signals has yet been found, leading some to question the feasibility of this approach. However, the ongoing advancements in telescope technology and signal processing capabilities offer renewed hope for future discoveries. The sensitivity of radio telescopes continues to improve, expanding the range of signals we can detect and increasing the probability of identifying even weak or distant transmissions.

Beyond the search for radio signals, SETI also encompasses the search for other technological signatures. This includes the detection of megastructures, such as Dyson spheres, hypothetical megastructures built around stars to harness their energy. The detection of such structures would undoubtedly signify the presence of a highly advanced civilization capable of manipulating stellar-scale energy resources. However, the identification of megastructures presents immense technological challenges. These structures would need to be sufficiently large and distinct from other astronomical objects to be detectable from Earth, which requires extremely powerful telescopes and sophisticated image processing techniques. Furthermore, the very definition of a technological signature remains a subject of ongoing debate. What constitutes unequivocal evidence of artificial origin? How do we distinguish natural phenomena from technological artifacts, particularly given our limited understanding of the vast diversity of potential extraterrestrial technologies?

Another approach in the search for extraterrestrial life involves the search for biosignatures – evidence of life, not necessarily intelligent life, elsewhere in the universe. This search focuses on the detection of molecules or other indicators that suggest the presence of biological processes. This approach has gained significant momentum with the discovery of numerous exoplanets, planets orbiting stars other than our sun. The detection of exoplanets has significantly broadened the scope of the search for life, providing a vastly expanded range of potential targets for investigation. Current research is focused on identifying exoplanets within the habitable zones of their stars, the regions where liquid water could exist on the planet's surface – a key requirement for life as we know it. However, the possibility of life existing outside of these habitable zones cannot be ruled out, particularly given the potential for subsurface oceans on icy moons or other environments with unique conditions that might support life.

The search for biosignatures involves the analysis of light from distant planets, looking for specific spectral features that might indicate the presence of atmospheric gases associated with life, such as oxygen, methane, or water vapor. This requires extremely powerful telescopes, such as the James Webb Space Telescope (JWST), which has the capability to analyze the atmospheric composition of exoplanets. However, the interpretation of such data presents significant challenges. The detection of biosignatures does not automatically prove the presence of life; other non-biological processes could potentially produce similar signatures. Careful analysis and interpretation are therefore crucial to differentiate genuine biosignatures from false positives. Furthermore, the vast distances involved make it extremely difficult to obtain high-resolution data on exoplanet atmospheres, limiting the accuracy of our analyses.

Beyond the detection of radio signals and biosignatures, the search for extraterrestrial life also involves a broader consideration of the physical and chemical conditions necessary for life to emerge and evolve. The study of extremophiles, organisms that thrive in extreme environments on Earth, such as hydrothermal vents or highly acidic environments, provides valuable insights into the potential for life to exist in seemingly inhospitable environments elsewhere in the universe.

These organisms demonstrate the remarkable resilience and adaptability of life, suggesting that life might be far more widespread and diverse than previously imagined. The understanding of extremophiles helps us broaden our search criteria, expanding the range of environments we consider potentially habitable and prompting further investigation of seemingly improbable locations for life.

Furthermore, the field of astrobiology, a multidisciplinary science combining astronomy, biology, chemistry, and geology, is playing an increasingly important role in the search for extraterrestrial life. Astrobiologists are studying the origin and evolution of life on Earth, using this knowledge to inform our search for life elsewhere. By understanding the processes that led to the emergence of life on Earth, we can better identify the conditions and factors that are likely to be necessary for life to arise elsewhere. This understanding can then be applied to the selection of potential targets for future research and the development of more sophisticated detection methods.

The search for extraterrestrial life is not merely a scientific endeavor; it is also a philosophical and existential quest. The discovery of extraterrestrial life, particularly intelligent life, would have profound implications for our understanding of our place in the universe, challenging our assumptions about the uniqueness of humanity and the nature of consciousness itself. The potential implications for society, science, and our understanding of our place in the universe are vast and far-reaching. This underscores the urgency and importance of continued research and exploration in this field, promising both scientific discovery and a deeper understanding of our own existence.

The search for extraterrestrial life is a long-term endeavor, requiring sustained investment in research, technology, and international collaboration. While the challenges are considerable, the potential rewards are immeasurable. The ongoing advancements in technology, the increasing number of known exoplanets, and the growing understanding of the conditions necessary for life to emerge, all point to the possibility of a future where we are no longer alone in our understanding of the universe. The quest to answer the question of whether we are alone remains one of humanity's greatest scientific and philosophical challenges, one that continues to inspire and drive our exploration of the cosmos. The continued pursuit of this question is a testament to our enduring curiosity and our innate desire to understand our place within the vastness of the universe.

The conventional search for extraterrestrial life, as compelling as it is, often operates under a tacit assumption: that these potential life forms exist within our shared three-dimensional spacetime continuum. However, the possibility that extraterrestrial life might originate from, or even interact with, other dimensions or realities opens up a profoundly different perspective. This section will explore the intriguing, albeit speculative, concept of extraterrestrial life engaging in interdimensional travel, venturing into realms far beyond our current comprehension.

The theoretical framework for such a possibility rests on several pillars. First, the burgeoning field of string theory and related concepts like M-theory suggest the existence of higher dimensions beyond our familiar four (three spatial and one temporal). These extra dimensions, often compactified or curled up at scales smaller than atoms, could potentially harbor entirely separate universes or "branes," each with its own physical laws and potentially, its own forms of life. If such parallel universes exist, and if beings within them possess sufficiently advanced technology – or perhaps even innate abilities – they could theoretically navigate these higher dimensions, achieving what we might term "interdimensional travel."

The challenge lies in conceptualizing the nature of such travel. Our understanding of physics is rooted in our four-dimensional experience. Movement through spacetime, as described by Einstein's theory of relativity, is governed by specific rules, with limitations imposed by the speed of light. Interdimensional travel, however, would likely transcend these limitations, possibly involving interactions with dimensions that are not directly accessible through our current understanding of spacetime. One could speculate on mechanisms that might facilitate such travel, perhaps involving manipulation of gravitational fields, quantum entanglement, or exotic forms of energy we have yet to discover. The very nature of these hypothetical methods might render them undetectable by our current scientific instruments.

Consider the possibility of "wormholes," hypothetical tunnels through spacetime predicted by Einstein's theory of general relativity. While mathematically plausible, their existence remains unproven. If wormholes do exist, they could theoretically connect distant regions of spacetime or even different universes, offering a pathway for interdimensional travel. However, the challenges associated with traversing a wormhole are immense. The immense gravitational forces at the entrance and exit of a wormhole could crush any spacecraft or traveler attempting to pass through. Furthermore, the stability of wormholes is highly questionable, with the potential for them to collapse before a traveler could pass through.

Another possibility centers on the concept of higher-dimensional beings. If life forms exist in higher dimensions, they might perceive our three-dimensional reality in a fundamentally different way, perhaps as a projection or a slice of their own higher-dimensional existence. From their perspective, interdimensional travel might not involve the same physical limitations as it would for beings confined to three dimensions. They might be able to interact with our reality at will, appearing and disappearing without violating any known physical laws, potentially explaining some accounts of unexplained phenomena often associated with paranormal activity.

The connection between interdimensional travel and the reported encounters with unidentified aerial phenomena (UAPs), often referred to as UFOs, presents another intriguing avenue of speculation. Some accounts of UAP sightings describe objects exhibiting impossible maneuvers, defying our understanding of aerodynamics and propulsion. Could such objects represent vehicles employing interdimensional technology? Could some UAP encounters be instances of extraterrestrial beings from other dimensions briefly interacting with our reality? It's crucial to approach such speculation with a critical and discerning eye, acknowledging the lack of definitive proof while remaining open to the possibility of explanations beyond our current scientific framework. The investigation of UAPs requires a rigorously scientific approach, carefully analyzing data and eliminating alternative explanations before considering more outlandish hypotheses.

Furthermore, the concept of "density" as discussed in previous chapters plays a crucial role in this discussion. Different densities could potentially correspond to different dimensional planes or realities. Beings inhabiting a denser realm might possess abilities that appear supernatural to beings from a less dense realm, such as the ability to teleport or manipulate matter in unconventional ways. This could provide a possible framework for interpreting accounts of paranormal abilities, potentially associating them with interactions between beings from different dimensional planes or realities. However, the challenge remains in establishing a concrete link between density, dimensions, and the observed phenomena.

The possibility of interdimensional travel also raises significant questions about communication. How could beings from different dimensional realities communicate with each other? Would their communication methods differ fundamentally from our own? Could they potentially access our thoughts or manipulate our perceptions directly? These are fundamental questions that challenge our current models of communication and understanding, requiring a deeper exploration of the nature of consciousness and information itself. Perhaps the alleged phenomenon of telepathy could be interpreted as a form of interdimensional communication.

The philosophical implications of extraterrestrial interdimensional travel are profound. If such beings exist, and if they possess advanced technologies or abilities far beyond our own, it would necessitate a reevaluation of our place in the cosmos. It would challenge our anthropocentric worldview, prompting us to reconsider our assumptions about the uniqueness of humanity and the nature of existence itself. The acceptance of such a possibility would be a giant leap in our understanding of reality, profoundly altering our scientific and philosophical perspectives.

The search for extraterrestrial life that engages in interdimensional travel, while deeply speculative, opens up vast and exciting avenues for investigation. It compels us to consider possibilities beyond our current scientific paradigms, pushing the boundaries of our knowledge and imagination. While concrete evidence remains elusive, the theoretical frameworks and observed anomalies warrant a continued exploration of this intriguing frontier. Ultimately, the quest to understand other realities and potential extraterrestrial interactions with these realms is a testament to human curiosity and our relentless pursuit of knowledge, even amidst the uncertainties and challenges of the unknown. The journey into the realm of interdimensional life may well be one of the most challenging and rewarding quests humanity has ever undertaken.

The intriguing possibility of interdimensional travel, explored in the previous section, naturally leads us to consider its potential connection to the enduring enigma of Unidentified Aerial Phenomena (UAPs), often colloquially referred to as UFOs. The sheer volume of reported UAP sightings, spanning decades and encompassing a diverse range of eyewitness accounts, presents a compelling challenge to conventional scientific explanation. While many sightings can be attributed to misidentified aircraft, weather phenomena, or hoaxes, a significant number remain unexplained, defying easy categorization and raising the tantalizing possibility of extraterrestrial, or even extra-dimensional, origins.

Many reported UAP encounters describe objects exhibiting remarkable capabilities, far exceeding the known limits of human aerospace technology. Reports frequently detail astonishing feats of maneuverability, including instantaneous acceleration, abrupt changes in direction at impossible speeds, and seemingly effortless suspension in mid-air – all defying the laws of Newtonian physics and our current understanding of aerodynamics and propulsion systems. These observed anomalies are not merely inconsistencies in observation; they fundamentally challenge our established scientific models, suggesting the possibility of propulsion systems operating on principles yet unknown to us.

One potential explanation, consistent with the framework of this book, lies in the possibility of interdimensional propulsion. If beings from other dimensions or realities can navigate the higher dimensions theorized in string theory or M-theory, their vehicles might utilize methods of propulsion that are invisible, or undetectable, to our three-dimensional sensors. These vehicles could seemingly materialize and dematerialize, traverse vast distances instantaneously, and execute maneuvers that appear to violate the laws of physics as we understand them, simply because they are operating outside the constraints of our three-dimensional spacetime.

The concept of density, explored in previous chapters, also plays a critical role in interpreting UAP sightings. If different densities correspond to different dimensional planes, then a UAP might be observed as a fleeting distortion in our reality, a momentary ripple in the fabric of spacetime as a higher-density object briefly intersects with our own. This could explain accounts of objects appearing and disappearing abruptly, shimmering or changing shape unexpectedly – fleeting glimpses into a reality operating at a different density level. The sheer unpredictability of many UAP encounters could stem from the inherent instability of such interdimensional crossings, subject to fluctuations in the interface between dimensions.

Furthermore, the accounts often include reports of unusual energy signatures or electromagnetic anomalies associated with UAP sightings. Some researchers have pointed to documented instances of electronic malfunctions near reported UAP events, including interference with radar systems, communication networks, and even the instrumentation of aircraft. This suggests a potential interaction with fields or energies that are outside our current comprehension, potentially linked to the hypothesized manipulation of higher-dimensional energy fields required for interdimensional travel.

The reports also suggest a wide range of observed UAP forms, from cigar-shaped objects to disc-like craft and highly irregular formations. This diversity could reflect the myriad of possible methods of interdimensional travel or perhaps the presence of diverse extra-dimensional life forms, each possessing unique technological capabilities. The lack of a consistent morphological pattern may actually strengthen the case for extra-dimensional origins, suggesting a multiplicity of approaches to navigating the higher dimensions rather than a unified technological standard.

It is crucial, however, to acknowledge the inherent challenges in investigating UAP reports. Eyewitness testimonies are often subjective and subject to biases and misinterpretations. The lack of physical evidence, such as recovered craft or tangible material samples, hinders scientific analysis. Furthermore, the deliberate fabrication of accounts, motivated by various factors ranging from self-promotion to political agendas, contributes to the overall difficulty of determining the veracity of specific reports. A rigorous approach to UAP investigation requires careful evaluation of each case, separating credible accounts from those lacking sufficient evidence or marred by obvious inconsistencies.

Nevertheless, the persistent nature of unexplained UAP sightings, combined with the compelling theoretical framework presented in this book, warrants continued investigation. The possibility of extra-dimensional encounters, while speculative, aligns with a growing body of evidence that suggests the universe is far more complex and multifaceted than our current scientific models allow for. By integrating cutting-edge theoretical physics with careful examination of reported anomalies, we might begin to unravel the mysteries surrounding UAPs and gain a deeper understanding of the potential realities that lie beyond our immediate perception. The apparent limitations of our current scientific instruments and conceptual frameworks underscore the need for a broader, more open-minded approach to this fascinating and perplexing subject.

The exploration of UAP sightings necessitates a multidisciplinary approach, bringing together expertise in physics, aerospace engineering, psychology, and even parapsychology. The potential connections between UAP encounters, paranormal experiences, and the hypotheses presented in this book emphasize the interconnectedness of what are often regarded as disparate fields of inquiry. A holistic investigation that transcends disciplinary boundaries is crucial to unraveling the complex tapestry of evidence.

While acknowledging the extraordinary nature of the claims associated with UAPs, we must also recognize the potential implications of a scientifically substantiated encounter with extra-dimensional entities. Such an event would irrevocably shift our understanding of the universe and our place within it. It would necessitate a reassessment of fundamental scientific principles and could revolutionize our philosophical understanding of consciousness, reality, and the nature of existence itself.

The scientific community faces a challenge in balancing rigorous skepticism with intellectual openness. Dismissing all unexplained UAP reports as mere misidentifications or hoaxes would be irresponsible, especially given the persistent nature of the phenomenon and the increasing sophistication of the observed anomalies. However, accepting every report at face value, without rigorous scrutiny and critical analysis, would equally be flawed. The path forward requires a nuanced and balanced approach: a commitment to thorough investigation, a willingness to consider unconventional hypotheses, and an acknowledgment that our current understanding of the universe may be incomplete.

The journey towards understanding UAPs, and the possibility of extra-dimensional interaction, is a testament to humanity's insatiable curiosity and our relentless pursuit of knowledge. It represents a bold challenge to our existing worldview, pushing the boundaries of our scientific understanding and expanding the horizons of our imagination. The quest for answers, even in the face of uncertainty and potential controversy, embodies the core spirit of scientific inquiry and reflects the enduring human yearning to comprehend the mysteries of the cosmos. The ultimate revelation, whether it involves a more complete understanding of our own three-dimensional reality or the discovery of other realities, promises a profound shift in our perception of the universe and our place within it. The ongoing exploration of UAPs, therefore, is not merely a scientific endeavor; it's a profound philosophical journey into the very fabric of existence.

Interpreting reported extraterrestrial encounters through the lens of our proposed multi-density universe model requires a shift in perspective. Current interpretations often rely on a single-density framework, assuming extraterrestrial life forms, if they exist, would interact with our reality in ways readily detectable by our current technological capabilities. This subsection challenges that assumption, proposing that variations in density might explain inconsistencies and ambiguities in reported encounters.

Consider the classic "flying saucer" archetype. Often described as metallic discs exhibiting impossible maneuvers, these reports frequently include accounts of intense light displays, electromagnetic disturbances, and even purported physical effects on witnesses. Within our model, these phenomena could be interpreted as manifestations of a higher-density object interacting with our lower-density reality. The seemingly impossible maneuvers might arise not from advanced propulsion systems as conventionally understood, but from the object's ability to manipulate density itself, effectively "sliding" between different dimensional planes. The intense light displays could be a byproduct of this density manipulation, a visible manifestation of energy fluxes as the object alters its dimensional position. Similarly, the reported electromagnetic disturbances could be attributed to the object's interaction with the electromagnetic fields of our reality, a consequence of its traversal across dimensional boundaries.

The reported physical effects on witnesses, such as feelings of nausea, disorientation, or even temporary paralysis, might not be the result of direct physical contact but rather a consequence of the disruption of energy fields at the interface between densities. Our bodies, being accustomed to a specific density, might react negatively to the energetic fluctuations caused by a high-density object's passage through our reality. This is analogous to the way our bodies react to extreme changes in pressure or temperature; the effect isn't direct physical contact but rather a disruption of the equilibrium of our internal systems.

Furthermore, many reported encounters involve alleged communication attempts, often described as telepathic or involving unintelligible symbols or sounds. Within our model, this could be explained by the difficulty of direct communication between beings operating at vastly different density levels. Direct verbal or visual communication might be impossible, requiring the utilization of subtle energy fields or subtle manipulation of consciousness, leading to the perceived telepathic or symbolic forms of communication. The inherent limitations of our sensory organs might restrict our ability to perceive the true nature of this communication, interpreting it as "telepathy" rather than a form of interdimensional signaling.

Consider the Roswell incident, a pivotal event in UFO lore. While many alternative explanations exist, a density-based interpretation offers a fresh perspective. If the object that crashed in Roswell was a higher-density entity that accidentally breached our reality, its physical characteristics might appear fundamentally different to our perception. The reported "metallic" properties might be a misinterpretation of a substance whose properties are fundamentally altered by the shift in density. The reported fragility of the materials might also be attributed to the instability of the object's existence within our lower-density plane, making the craft inherently prone to disintegration upon disruption of its dimensional integrity.

The apparent lack of tangible evidence in many cases is also readily explained within our framework. If a higher-density object traverses our reality briefly, it might leave behind no trace aside from fleeting distortions in energy fields. Our current scientific instruments, designed to detect phenomena within our known density range, might be insufficient to detect these subtleties. This doesn't negate the experience; it simply highlights the limitations of our current observation tools and theoretical models.

The inconsistencies in eyewitness accounts, a common criticism leveled against the validity of many reports, can also be explained by the inherent variability of inter-dimensional crossings. The object's appearance, behavior, and even the experienced effects on the witnesses might change depending on the stability of the density interface at the time of the encounter. The fluctuating nature of the dimensional barrier could cause variations in the object's perceived form, leading to contradictory accounts from different observers witnessing the same event.

A significant challenge in interpreting these encounters is the potential for psychological factors to influence eyewitness accounts. Stress, fear, and the sheer strangeness of the situation can distort perception and memory. However, it is important to acknowledge that even seemingly improbable accounts can contain kernel of truth that must be carefully considered. A rigorous approach necessitates evaluating all elements of the encounter, acknowledging both the limitations of human perception and the potential of phenomena beyond current scientific understanding.

The concept of time also requires careful consideration. Many accounts describe experiences of time dilation, where the perceived passage of time during the encounter differed significantly from the measured passage of time afterward. This aligns with our multi-density model, where time might not be a constant across different density planes. A higher-density realm might experience time differently, causing a temporal discrepancy between the experience of the witness and the external observer.

Beyond the specific case studies, our model provides a unifying framework for interpreting various paranormal phenomena often associated with UFO encounters. Reports of poltergeist activity, levitation, and telekinesis could all be interpreted as manifestations of energy fluxes or dimensional instability occurring at the interface between density levels. This interconnection underscores the need for a more holistic approach, integrating parapsychology with theoretical physics to construct a more complete understanding of the interplay between different dimensional planes.

Moving forward, a crucial step is to develop new methods of detecting and measuring density variations. Current scientific instruments, optimized for three-dimensional observations, might be inadequate for detecting higher-density phenomena. This necessitates the development of advanced sensors capable of detecting subtle changes in energy fields or other indicators of density shifts. This research could involve exploring technologies such as advanced gravitational wave detectors, highly sensitive electromagnetic sensors, and perhaps even devices based on yet-unknown principles that could directly interact with higher dimensional energy fields.

Furthermore, a multidisciplinary approach, integrating physics, psychology, and parapsychology, is crucial. Combining the perspectives of these fields can help filter subjective accounts and identify objective data. Rigorous statistical analysis, applying techniques developed to study other anomalous phenomena, will help distinguish genuine events from those attributable to chance, misinterpretation or outright fabrication.

The investigation of extraterrestrial encounters should not be considered a fringe pursuit but rather a critical area of scientific inquiry that could profoundly shape our understanding of the universe. The potential for discovering other realities, other forms of life, and other fundamental laws of physics demands open-minded investigation, rigorous methodology, and a willingness to challenge our current scientific paradigms. The answers may lie beyond our current understanding, but the pursuit of these answers is crucial for pushing the boundaries of human knowledge and our understanding of the universe. The exploration of these enigmas is not merely an intellectual exercise; it is a profound journey into the very fabric of existence itself.

The implications of establishing contact with extraterrestrial life are profound and multifaceted, extending far beyond the simple exchange of greetings. Our current understanding of extraterrestrial life, largely shaped by science fiction, often portrays encounters as straightforward interactions between technologically advanced civilizations. However, the possibility of contact with beings inhabiting different density planes, as proposed in our multi-density universe model, necessitates a far more nuanced and cautious approach.

One of the most significant implications revolves around the potential for technological disparity. If we encounter a civilization significantly more advanced than our own, the power imbalance could be overwhelming. This disparity wouldn't necessarily be limited to technological superiority in weaponry or energy production. It could encompass control over fundamental forces of nature, including manipulation of density itself. Such a civilization might possess the ability to alter our reality in ways we cannot comprehend, posing existential risks to our species. Imagine a scenario where a higher-density civilization, seeking resources or expansion, views our reality as merely a lower-density substrate, readily manipulated for their purposes. The consequences could be catastrophic, ranging from subtle alterations of our physical laws to outright annihilation.

Conversely, the less likely but still possible scenario of encountering a less technologically advanced civilization presents its own ethical dilemmas. We are currently grappling with the complexities of interstellar travel and the potential impact on other planetary systems. The ethical considerations inherent in encountering a less advanced civilization are far more complex than merely avoiding contamination. Should we intervene to help such a civilization develop, or should we observe from a distance, respecting their right to autonomy and independent development? This delicate balance between benevolence and interference requires careful ethical consideration, establishing a clear framework for interstellar interaction before contact is ever made. Such a framework would need to encompass not just technological and scientific considerations but also philosophical, ethical, and even spiritual aspects of potential interactions.

The nature of communication itself presents another significant hurdle. As discussed previously, communication between entities operating at vastly different density levels could be extraordinarily challenging. Even if a shared communication medium is found, the conceptual frameworks and understanding of reality itself might be so different that mutual understanding becomes virtually impossible. The very act of attempting communication could have unforeseen consequences, potentially resulting in unintended misunderstandings or disruptions to the fabric of reality. Our model suggests that communication might necessitate advanced forms of energetic manipulation, operating on frequencies and principles beyond our current comprehension. The possibility of misinterpreting signals, or even initiating unintentional interactions with unintended consequences, emphasizes the importance of caution and careful planning in any approach to contact.

Beyond the immediate implications of contact, the discovery of extraterrestrial life, particularly life inhabiting different density planes, would revolutionize our understanding of the universe. It would necessitate a complete re-evaluation of our scientific models, challenging long-held assumptions about the nature of reality, physics, and even consciousness. The implications for cosmology, astrophysics, and even philosophy would be staggering, potentially opening up new avenues of scientific inquiry and challenging the foundations of human knowledge.

The sociological impact of such a discovery would also be transformative. Human societies have always been shaped by their understanding of the universe and their place within it. The discovery of extraterrestrial life would inevitably change our self-perception and redefine our position in the cosmos. Existing belief systems, cultural narratives, and social structures could be challenged and reshaped in profound ways. The possibility of multiple forms of intelligent life, each with unique characteristics and potentially vastly different understanding of reality, could lead to fundamental changes in our philosophical perspectives, ethical frameworks, and even spiritual beliefs. The consequences could range from increased cooperation and global unity to heightened tensions and conflict, depending on the nature of the contact and its implications for humanity.

Furthermore, the economic and political ramifications of contact are potentially immense. The discovery of new resources, technologies, or even trade partners could reshape global economies, power dynamics, and political alliances. However, the distribution of such benefits, as well as potential threats, could create immense social and political instability. Careful international cooperation and a global framework for managing such interactions are essential to minimize potential conflicts and to maximize the benefits of contact.

Finally, the psychological impact of contact must be considered. The sudden realization that humanity is not alone in the universe, that there are other intelligent beings, possibly even beings inhabiting realities beyond our comprehension, could have a significant psychological impact on individuals and societies. Such a profound revelation could lead to various reactions, ranging from intense awe and wonder to fear, anxiety, and even social upheaval. A well-planned and carefully executed strategy for managing the psychological implications of contact is essential to mitigate potential negative impacts and to foster a sense of optimism and unity in the face of this momentous event.

The potential implications of extraterrestrial contact are vast and complex, touching upon every aspect of human existence. From the technological and ethical considerations to the sociological, political, economic, and psychological impacts, we must approach the possibility of contact with a combination of scientific rigor, ethical responsibility, and a deep understanding of the potential ramifications. Preparing for such a monumental event necessitates a global, multi-disciplinary approach, ensuring that humanity is ready to confront this challenge, not with fear, but with the wisdom, resilience, and foresight required to navigate this pivotal moment in our history. The discovery of extraterrestrial life, particularly life existing in different density planes, isn't merely a scientific event; it's a watershed moment that will forever redefine our understanding of ourselves and our place in the cosmos. The careful and deliberate consideration of these implications is paramount, not as a deterrent to exploration but as a critical step in ensuring a safe and successful future for humanity in a universe vastly more complex and wondrous than we currently imagine.

7

The Physics of Other Realities

The exploration of extraterrestrial life, particularly life potentially residing in different density planes, naturally leads us to consider the profound implications of quantum physics and its intriguing connection to the concept of parallel universes. The very fabric of reality, as understood through the lens of quantum mechanics, suggests a universe far more complex and multifaceted than our classical intuition allows. This complexity opens the door to the tantalizing possibility of multiple realities, coexisting alongside our own, perhaps even interacting with it in subtle and unexpected ways.

At the heart of this connection lies the principle of quantum superposition. In the quantum realm, unlike the macroscopic world we experience daily, particles don't exist in definite states until measured. Instead, they exist in a superposition of all possible states simultaneously. Think of Schrödinger's cat, the famous thought experiment illustrating this paradox: until the box is opened and the cat's state is observed, the cat is both alive and dead simultaneously. This seemingly paradoxical behavior challenges our classical understanding of reality, suggesting that the universe, at its most fundamental level, exists in a state of probabilistic potentiality.

This probabilistic nature of quantum mechanics is not confined to individual particles; it extends to the entire universe. Some interpretations of quantum mechanics suggest that every quantum measurement causes the universe to split, creating multiple branching realities, each representing a different possible outcome of the measurement. In this "many-worlds" interpretation, every decision, every event, results in a branching of reality, creating an infinite number of parallel universes, each with its own unique timeline and history. Our reality, then, is merely one branch on an infinitely branching tree of possibilities.

The implications of this are staggering. If parallel universes exist, they might harbor vastly different physical laws, constants, and even fundamental particles. These universes might be indistinguishable from our own, differing only in seemingly minor details, or they might be radically different, governed by entirely different physics and containing forms of matter and energy beyond our current comprehension. The sheer scale and variety of these parallel universes are virtually limitless, potentially encompassing every conceivable possibility.

Quantum entanglement further strengthens the case for interconnected parallel universes. Entanglement is a phenomenon where two or more particles become linked in such a way that their fates are intertwined, regardless of the distance separating them. Measuring the state of one entangled particle instantly determines the state of the other, even if they are light-years apart. This instantaneous correlation seems to violate the principles of locality and causality, suggesting a deeper connection between these particles than our current understanding of space and time allows.

Some physicists propose that entangled particles are not merely linked but are actually manifestations of the same underlying reality, residing in different parts of a larger multiverse. The instantaneous correlation between entangled particles might be explained by their shared existence in a higher-dimensional space, where distance and time as we perceive them are irrelevant. This hypothesis suggests that parallel universes aren't completely isolated from each other but are interconnected through quantum entanglement, allowing for subtle interactions and information exchange between them.

The connection between quantum physics and the existence of parallel universes is not universally accepted, and many alternative interpretations of quantum mechanics exist. However, the possibilities raised by these interpretations are too intriguing to ignore. The probabilistic nature of the quantum realm, combined with the mysterious phenomenon of entanglement, suggests a universe far richer and more complex than our classical intuition can grasp.

The implications for parapsychology are equally profound. Phenomena such as precognition, telepathy, and psychokinesis, often dismissed as unscientific, might find a plausible explanation within the framework of a multiverse. If parallel universes exist and are interconnected through quantum entanglement, it's conceivable that information could leak between them, allowing for experiences such as precognitive dreams or telepathic communication. These phenomena, typically considered paranormal, might simply be manifestations of quantum interactions between our reality and other parallel universes.

Furthermore, the concept of parallel universes offers a new perspective on the nature of consciousness. If our consciousness is not confined to a single universe but extends across multiple realities, it could explain some of the perplexing aspects of near-death experiences, out-of-body experiences, and other altered states of consciousness. The experience of being "out of body," for example, could be interpreted as a temporary excursion into a parallel universe, a momentary glimpse into a different aspect of our own consciousness.

It's crucial to emphasize that these are theoretical speculations, and the existence of parallel universes remains unproven. However, the theoretical framework of quantum mechanics, with its inherent probabilistic nature and the phenomenon of entanglement, provides a foundation for the serious consideration of these possibilities. The exploration of these ideas necessitates a blurring of the boundaries between established scientific disciplines and areas often considered outside the realm of mainstream science. The careful, rigorous, and open-minded investigation of these theories could lead to breakthroughs in our understanding of reality, consciousness, and our place in the cosmos. It could be the key to unlocking a more comprehensive understanding of the universe, potentially revealing dimensions and realities beyond our current wildest imaginations.

The search for evidence of parallel universes requires a multidisciplinary approach, combining the tools and techniques of quantum physics, cosmology, and parapsychology. This requires a paradigm shift, a willingness to explore unconventional avenues of research and to reconsider our preconceived notions about the nature of reality. The investigation may involve searching for subtle quantum correlations that could indicate interactions between parallel universes, analyzing data from near-death experiences and other altered states of consciousness, or developing new theoretical models that incorporate the concept of multiple realities.

The pursuit of such knowledge is not merely an academic exercise; it has profound philosophical, ethical, and spiritual implications. The discovery of parallel universes would fundamentally alter our understanding of ourselves, our place in the universe, and the nature of reality itself. It could challenge our deepest beliefs about consciousness, free will, and the possibility of life beyond our own. The potential rewards of this quest are immense, potentially leading to breakthroughs in our understanding of fundamental physics, the nature of consciousness, and the very fabric of reality. The journey may be fraught with challenges, but the potential discoveries are worth the risk. The exploration of parallel universes is not just a scientific pursuit; it is a journey into the heart of existence itself, a quest to unravel the deepest mysteries of the cosmos and our place within it. This pursuit necessitates a profound reevaluation of our methodologies and assumptions, acknowledging the inherent limitations of our current understanding and embracing the unknown with intellectual humility and courage. The rewards of such an undertaking could redefine our understanding of reality itself, forever altering the trajectory of human knowledge and understanding.

Building upon the probabilistic nature of quantum mechanics and the enigma of entanglement, we can delve deeper into the theoretical frameworks that suggest the existence of other realities. String theory and braneworld scenarios, while highly abstract and still largely theoretical, offer compelling possibilities that align with the broader concept of a multiverse. These theories propose not only the existence of parallel universes, but also the existence of extra spatial dimensions beyond our familiar three.

String theory, perhaps the most ambitious attempt at a "theory of everything," postulates that the fundamental constituents of the universe are not point-like particles, but rather tiny, vibrating strings. These strings are incredibly small, far smaller than even the most fundamental particles we've observed. Their vibrations, or modes of oscillation, determine the properties of the particles they constitute. This elegant simplicity offers a potential explanation for the seemingly disparate forces of nature, unifying gravity with the other fundamental forces.

However, string theory's elegance comes at a cost. To make the mathematics work, string theory requires the existence of extra spatial dimensions, typically curled up and compactified at an incredibly small scale, imperceptible to our current observational capabilities. These extra dimensions are not merely theoretical constructs; their existence is a mathematical necessity within the framework of string theory. The exact number of these dimensions and the manner in which they are compactified are still subjects of intense research and debate, with different versions of string theory proposing different scenarios.

The concept of braneworlds emerges as a natural extension of string theory and higher-dimensional models. In this scenario, our observable universe, with its four dimensions (three spatial and one temporal), is not the entirety of existence but is rather a three-dimensional "brane" embedded within a higher-dimensional spacetime, sometimes referred to as the "bulk." Other branes, representing other universes, could also exist within this higher-dimensional bulk, separated from our brane by the extra dimensions.

The implications of this model are profound. The gravity we experience is a manifestation of the gravitational field propagating through the bulk, while other forces, like electromagnetism and the strong and weak nuclear forces, are confined to our brane. This explains why gravity is so much weaker than the other fundamental forces; its influence is spread over a larger, higher-dimensional space. The existence of other branes, potentially with different physical laws and constants, opens the door to a vast multiverse, each brane representing a distinct universe with its unique characteristics.

Interactions between branes could be rare, or even impossible, depending on the specifics of the model. However, some scenarios suggest the possibility of brane collisions, which could have catastrophic consequences for our universe. Other interactions could be more subtle, perhaps manifested as unusual gravitational anomalies or even quantum entanglement effects that seem to bridge the gap between different universes. These hypothetical interactions could provide a potential pathway for testing the braneworld hypothesis.

The difficulty in directly observing or confirming these extra dimensions and other branes is significant. Our current technology is limited in its ability to probe such small scales and high energies. The compactification of these extra dimensions means they are curled up so tightly that they remain effectively invisible to our current instruments. However, indirect evidence for their existence might be detectable through subtle effects on gravity or through the observation of high-energy cosmic phenomena.

The quest for indirect evidence is a major focus of ongoing research in cosmology. The precise measurement of the cosmic microwave background radiation, the afterglow of the Big Bang, is one promising avenue of investigation. Deviations from theoretical predictions in the CMB could hint at the existence of extra dimensions or other unforeseen cosmic features associated with braneworld scenarios. Similarly, searches for gravitational waves, ripples in spacetime, could reveal evidence of brane collisions or other high-energy events occurring in the bulk.

The search for experimental evidence of string theory and braneworlds is challenging, but not impossible. The next generation of particle accelerators, such as the proposed Future Circular Collider, could potentially probe energies high enough to detect signatures of string-theoretic phenomena or the effects of extra dimensions. Advances in gravitational wave detection technology could also play a crucial role, enabling us to identify subtle gravitational perturbations indicative of brane interactions.

Furthermore, the theoretical implications of string theory and braneworlds are not limited to physics alone. They have far-reaching implications for our understanding of cosmology, including the nature of dark matter and dark energy. Dark matter, the mysterious substance making up a significant portion of the universe's mass, could potentially be related to particles or fields residing in the bulk, influencing our universe through gravitational interactions. Similarly, dark energy, the force driving the accelerated expansion of the universe, might have its origin in the dynamics of higher-dimensional spacetime.

The exploration of string theory and braneworlds involves tackling complex mathematical frameworks, and the development of advanced computational tools is crucial. String theory calculations often involve intricate mathematical techniques that push the boundaries of our current computational capabilities. The development of novel algorithms and advanced computing infrastructure is vital for progressing in our understanding of these complex theoretical structures.

Beyond the immediate scientific pursuit, the implications of string theory and braneworlds touch upon the philosophical and metaphysical. If these theories accurately reflect reality, they dramatically alter our perception of the universe and our place within it. Our three-dimensional world becomes a tiny slice of a vastly larger, higher-dimensional reality. The possibility of other branes, each potentially harbouring unique universes with different physical laws and constants, raises profound questions about the nature of existence, the uniqueness of our universe, and the potential for life beyond our immediate realm.

While string theory and braneworlds remain theoretical frameworks, their exploration drives us to develop new mathematical tools, experimental techniques, and modes of thinking. The journey toward a deeper understanding of these models is a crucial step in our ongoing quest to unravel the mysteries of the universe. The inherent challenges involved are substantial, but the potential rewards – a profound expansion of our understanding of reality and our place within it – are equally immense, motivating continued research in this fascinating and challenging realm of theoretical physics. This pursuit represents a critical frontier in scientific exploration, pushing the boundaries of our current understanding and inviting further investigation into realms hitherto unimaginable. The journey promises to be long and arduous, but the possibility of uncovering the deepest secrets of existence is a compelling incentive for continued effort in these exciting fields.

Loop quantum gravity (LQG) offers a radically different perspective on the universe's structure compared to the more familiar general relativity. Instead of treating spacetime as a smooth, continuous entity as Einstein's theory does, LQG proposes that spacetime itself is quantized. This means that spacetime is not infinitely divisible; rather, it's composed of fundamental, discrete units, analogous to the way energy is quantized into photons or matter into quarks. These fundamental units of spacetime are often referred to as "loops" or "spin networks," giving the theory its name.

Imagine spacetime as a vast, intricate network woven from these fundamental loops. Each loop represents a fundamental unit of area or volume, and the connections between these loops define the geometry of spacetime. This granular structure is not readily apparent at the macroscopic level, where spacetime appears smooth and continuous. However, at extremely small scales, near the Planck length (approximately 10^-35 meters), the discrete nature of spacetime would become dominant. This is where the quantum effects become overwhelmingly significant and the continuous picture breaks down.

The implications of LQG for the universe's structure are profound. The quantized nature of spacetime means that at the smallest scales, there are no singularities, the points of infinite density predicted by general relativity in the context of black holes and the Big Bang. Instead, LQG suggests that these singularities are replaced by extremely dense but finite regions of spacetime. This could fundamentally alter our understanding of the origins of the universe and the final fate of black holes.

Furthermore, LQG introduces the concept of a "quantum bounce," offering an alternative to the Big Bang singularity. In this scenario, the universe didn't begin from a singular point of infinite density but instead emerged from a previous contracting phase, reaching a state of maximum density before bouncing and expanding into the universe we observe today.

This cyclical model of the universe, constantly contracting and expanding, eliminates the need for a singular beginning, offering a more complete and perhaps more elegant cosmological picture.

The implications of LQG extend beyond the Big Bang and black holes. It could also resolve some of the paradoxes associated with black holes, such as the information paradox. This paradox arises from the seeming loss of information when matter falls into a black hole, a violation of fundamental principles of quantum mechanics. LQG's discrete structure of spacetime might provide a mechanism for preserving this information, potentially resolving the paradox.

Moreover, LQG offers a potential framework for unifying gravity with the other fundamental forces of nature. While string theory attempts this unification by postulating tiny vibrating strings as fundamental building blocks, LQG tackles the problem from a different angle, by quantizing spacetime itself. The success of such a unification would represent a monumental achievement in theoretical physics, offering a complete description of all the fundamental interactions in the universe.

The connection between LQG and the multiverse hypothesis is less direct than with string theory but equally intriguing. The discrete nature of spacetime in LQG suggests a potential for a vast number of possible spacetime configurations. Each configuration, determined by the arrangement and connections of the fundamental loops, could represent a distinct universe within a broader multiverse. This multiverse wouldn't necessarily be composed of parallel universes existing in separate spatial regions, but rather different possible states of spacetime itself, each representing a self-contained universe with its own distinct physical laws and properties.

The mathematical complexity of LQG presents a formidable challenge. Calculations involving spin networks are computationally intensive, requiring the development of sophisticated mathematical tools and computational techniques. However, recent advances in this area are beginning to yield insights into the behavior of these intricate quantum spacetime structures, paving the way for a deeper understanding of the theory's implications for cosmology.

The theoretical landscape is not without its uncertainties. LQG is still a work in progress, and significant challenges remain in testing its predictions against observational data. The incredibly small scales at which the quantum nature of spacetime becomes evident make direct experimental verification nearly impossible with current technology. However, indirect evidence could potentially be found through subtle effects on the cosmic microwave background radiation, gravitational waves, or other cosmological phenomena.

Nevertheless, LQG's potential to resolve some of the most fundamental questions in cosmology and theoretical physics makes it a compelling area of research. The possibility of a quantized spacetime, a quantum bounce, and a multiverse generated from the multitude of possible spacetime configurations are truly mind-boggling and present a dramatically different picture of reality compared to our current understanding.

The ongoing efforts to develop and test LQG are not merely academic pursuits. They are attempts to understand the very fabric of reality, to probe the deepest mysteries of the universe, and potentially to uncover a far more complex and expansive reality than we could have ever imagined. The ramifications of such a profound shift in our understanding extend far beyond physics, touching upon our philosophical and metaphysical conceptions of existence. The search for a unified theory of everything, a theory that explains the universe's fundamental workings from the smallest to the largest scales, is a quest for understanding our place in the cosmos and the nature of reality itself.

The potential connection between LQG and parapsychology is, admittedly, highly speculative. However, if we accept the possibility of a multiverse composed of numerous spacetime configurations, some with drastically different physical laws, it's conceivable that certain phenomena currently categorized as "paranormal" could find a rational explanation within this expanded framework. For example, quantum entanglement, already a perplexing phenomenon within our own universe, could be amplified or manifested in unexpected ways in universes with different physical constants, potentially leading to seemingly inexplicable events. The possibility of interactions between different universes, even extremely rare ones, could also offer alternative explanations for seemingly impossible occurrences, such as telepathy or precognition.

This is not to say that LQG directly explains or validates all paranormal phenomena. Such interpretations remain highly speculative and require far more investigation. However, the theoretical framework of LQG, with its implications for a far more complex and potentially interconnected multiverse, opens the door to considering explanations that fall outside our current paradigms. It provides a scientific foundation for exploring unconventional phenomena with an open mind, even as it challenges the very limits of our current scientific understanding.

The journey toward a unified theory encompassing both quantum mechanics and general relativity remains a central focus of theoretical physics. The path towards verifying LQG's predictions and exploring its implications for cosmology and other seemingly unrelated fields like parapsychology promises to be long and fraught with challenges. However, the immense potential rewards—a deeper understanding of the universe, its structure, and its mysteries—make it a compelling and necessary pursuit. The very act of grappling with such abstract and intricate concepts pushes the boundaries of human thought, forcing us to constantly reevaluate our assumptions and expand our conceptions of reality. It is this constant intellectual exploration that fuels the engine of scientific progress, driving us toward ever-deeper truths about the universe and our place within it. The unification of quantum gravity with cosmology represents not just a scientific challenge but also a significant step towards a more complete and nuanced understanding of existence itself. This endeavor, regardless of its final outcome, will undoubtedly enrich our knowledge and shape our future understanding of reality for generations to come.

Building upon the foundational concepts of loop quantum gravity (LQG) and its implications for a potentially vast multiverse, we now delve into the mathematical models used to describe these other realities. These models, while often highly abstract and complex, are crucial for making testable predictions and for evaluating the plausibility of different multiverse scenarios. They bridge the gap between theoretical speculation and potential empirical verification, however challenging that verification might be.

One prominent approach is the application of higher-dimensional geometries. While our everyday experience is confined to three spatial dimensions and one time dimension, many theoretical frameworks suggest the existence of additional spatial dimensions, curled up or compactified at scales far smaller than those accessible to current experimental techniques. String theory, for instance, requires the existence of ten or eleven dimensions for its mathematical consistency. These extra dimensions are not directly observable in our four-dimensional spacetime, but their presence might influence the physical laws and constants that govern our universe.

Mathematical models incorporating these extra dimensions typically involve complex mathematical structures such as Calabi-Yau manifolds. These are six-dimensional spaces with highly intricate geometries, capable of producing a wide variety of physical properties. Different Calabi-Yau manifolds correspond to different possible compactifications of the extra dimensions, each potentially leading to a distinct universe with its own unique set of physical laws and constants. The sheer number of possible Calabi-Yau manifolds is astronomically large, suggesting the possibility of an enormous multiverse encompassing an almost infinite variety of universes.

The challenge lies in predicting which specific Calabi-Yau manifold, or combination of manifolds, corresponds to our own universe. This is a complex mathematical problem, requiring sophisticated computational techniques and a deep understanding of both string theory and differential geometry. Progress has been made, but a definitive solution remains elusive. Moreover, experimentally verifying the existence of these extra dimensions is an immense challenge, requiring technologies far beyond our current capabilities.

Beyond higher-dimensional geometries, other mathematical models explore the concept of brane worlds. This theory posits that our universe is confined to a three-dimensional "brane" embedded within a higher-dimensional spacetime, called the "bulk." Other branes could exist within the same bulk, representing parallel universes. These branes could interact with each other through gravitational forces, or perhaps through other yet-undiscovered interactions.

The mathematical description of brane worlds involves general relativity in higher dimensions, a significant challenge given the inherent complexities of Einstein's theory in even four dimensions. Models have been developed to describe the dynamics of branes, their potential interactions, and their influence on the properties of our own universe. However, many of these models rely on assumptions and approximations that need further refinement and validation. Again, experimental verification remains a formidable hurdle.

Furthermore, the landscape of possibilities expands when we consider the concept of different fundamental constants. In our universe, fundamental constants such as the speed of light, the gravitational constant, and the Planck constant have specific values that determine the properties of matter, energy, and spacetime. But other universes might exist with different values of these constants. The mathematical implications of this are profound, as a change in even a single constant could dramatically alter the characteristics of a universe. Some universes might be devoid of stars and galaxies, others might be unstable and collapse quickly, and yet others might be teeming with life forms unimaginable to us.

Mathematical models exploring this realm often involve varying the parameters of fundamental equations in physics, such as the Standard Model of particle physics or general relativity. This allows for the creation of simulated universes with different fundamental constants and the exploration of their resulting properties. These simulations can provide valuable insights into the range of possibilities encompassed by a multiverse, although they are inevitably limited by the assumptions and approximations inherent in the models themselves.

The mathematical modeling of these other realities extends beyond the previously discussed topics and encompasses numerous other approaches. For instance, some models explore the possibility of universes with different topological structures or different numbers of spatial dimensions. Others investigate universes governed by different laws of physics altogether, deviating from our own in profound ways. Each of these models introduces its unique set of mathematical challenges and conceptual complexities.

It's important to emphasize that many of these models are highly speculative, resting on theoretical frameworks that haven't been fully tested experimentally. The sheer mathematical complexity often necessitates simplifications and approximations, which can limit the reliability of their predictions. Nonetheless, these models provide invaluable tools for exploring the theoretical possibilities of other realities, guiding the development of future experiments, and pushing the boundaries of our understanding of the cosmos.

Connecting these mathematical models back to the realm of parapsychology requires a careful and nuanced approach. While it's tempting to invoke the multiverse hypothesis as a potential explanation for seemingly inexplicable paranormal phenomena, the evidence for such phenomena remains tenuous. Moreover, even if the multiverse exists, the mechanisms through which such interactions might occur would require substantial theoretical development.

However, the sheer possibility of other realities, described by these complex mathematical models, opens new avenues for exploring the foundations of consciousness, the nature of reality, and the relationship between the physical world and subjective experience. The very act of constructing and analyzing these models forces us to confront the limits of our current scientific understanding, pushing us beyond established paradigms and encouraging us to imagine possibilities previously considered inconceivable. The quest to understand the universe, even in the context of these highly speculative models, is a testament to the boundless curiosity and relentless pursuit of knowledge that define human nature. The exploration of other realities, guided by mathematics and fueled by imagination, leads us to a deeper understanding of ourselves and our place within the vast and mysterious cosmos.

The challenge lies not just in developing sophisticated mathematical models, but also in devising experimental tests that could potentially confirm or refute their predictions.

This might involve searching for subtle anomalies in the cosmic microwave background radiation, detecting faint gravitational signatures from interacting branes, or developing entirely new experimental techniques capable of probing the structure of spacetime at the Planck scale. The path forward requires a synergy between theoretical innovation and technological advancement. The journey toward understanding the physics of other realities is a long and arduous one, but the potential rewards – a profound shift in our understanding of the universe and our place within it – make it a journey well worth undertaking. The implications are not just scientific, but also philosophical and potentially even theological, forcing us to question our assumptions about the nature of reality, consciousness, and existence itself. The exploration continues, guided by the power of mathematical models and the unyielding spirit of scientific inquiry.

The theoretical frameworks discussed thus far, while providing compelling mathematical models for the existence of other realities, ultimately remain speculative. Their power lies in their ability to generate testable hypotheses, but the practical challenges of verifying these hypotheses are immense. The very nature of "other realities," by definition, implies their elusiveness, existing perhaps beyond the reach of our current scientific instruments and methodologies. Yet, the pursuit of scientific understanding compels us to explore potential avenues for empirical investigation.

One promising avenue lies in the realm of cosmology. The cosmic microwave background (CMB), the afterglow of the Big Bang, holds within its subtle fluctuations a wealth of information about the early universe. Anomalies in the CMB, deviations from the expected patterns of temperature and polarization, could potentially provide evidence for interactions with other universes or different regions of a multiverse. While the current CMB data is largely consistent with the standard cosmological model, future, higher-resolution observations might reveal subtle signatures of interactions that were previously undetectable. The search for such anomalies requires increasingly sensitive instruments and sophisticated data analysis techniques, pushing the boundaries of observational cosmology. The potential payoff, however, is enormous – direct observational evidence for the existence of other realities.

Another area of potential investigation focuses on gravitational waves. These ripples in spacetime, predicted by Einstein's theory of general relativity and now directly observed, provide a unique window into some of the most energetic events in the universe. If our universe interacts with other universes or branes, this interaction could manifest as subtle perturbations in the gravitational wave signal. The detection of such perturbations, while extremely challenging given the inherent weakness of gravitational waves and the potential masking effects of other astrophysical sources, could offer another avenue for confirming the existence of other realities. Advances in gravitational wave detectors, with increased sensitivity and improved data analysis techniques, might enable us to explore this exciting possibility in the coming decades.

Beyond cosmology, particle physics experiments might also play a role. High-energy particle collisions, such as those conducted at the Large Hadron Collider (LHC), could potentially produce particles that "leak" into other dimensions or interact with other universes. The detection of such particles would provide unambiguous evidence for the existence of extra dimensions or parallel universes. However, the theoretical predictions for the production and detection of such particles are often highly model-dependent, and the experimental signatures might be extremely subtle or difficult to distinguish from background noise. Moreover, the energies required to probe these hypothetical interactions might far exceed the capabilities of even the most powerful future colliders.

Furthermore, the exploration of other realities might necessitate the development of entirely new experimental technologies. Our current understanding of physics might be incomplete, and new paradigms might be required to access and study other realities. This could entail developing new sensors capable of detecting subtle interactions with other dimensions or universes, or creating entirely new experimental setups that bypass the limitations of our current technologies. The search for other realities could be the catalyst for revolutionary breakthroughs in physics, engineering, and related fields.

The challenges associated with testing hypotheses about other realities are not merely technical; they are also deeply philosophical. The very concept of an "other reality" raises fundamental questions about the nature of observation, measurement, and the limits of scientific inquiry. If other realities are truly separate from our own, how can we hope to interact with them, or even detect their presence? How do we define and measure the properties of a reality that fundamentally differs from our own? These are complex questions with no easy answers, requiring careful philosophical analysis alongside the scientific investigation. The epistemological challenges are intertwined with the methodological ones, necessitating a holistic approach that integrates both scientific rigor and philosophical reflection.

The possibility of other realities also raises ethical and societal considerations. If we were to discover evidence of other intelligent life forms, either within our own universe or in a parallel universe, how would we approach such an encounter? What ethical principles would guide our interactions, and how would we ensure the well-being of both ourselves and any other intelligent beings we might discover? These are profound questions, far-reaching in their implications, requiring careful consideration by scientists, philosophers, policymakers, and the public at large.

Furthermore, the exploration of other realities has the potential to fundamentally reshape our understanding of consciousness. If other realities exist, what implications does this have for the nature of consciousness? Could consciousness exist independently of the physical world, perhaps bridging different realities or even existing across multiple universes? These are fundamental questions that lie at the intersection of physics, philosophy, and neuroscience. The investigation of other realities necessitates a multidisciplinary approach, bringing together expertise from various fields to tackle these profound and interconnected questions.

The quest for evidence of other realities is a long-term endeavor, requiring sustained investment in research, development of new technologies, and a commitment to interdisciplinary collaboration. The journey is likely to be filled with unexpected challenges and setbacks, but the potential rewards—a revolutionary shift in our understanding of the cosmos, ourselves, and our place within the vast expanse of existence—are immense. It is a testament to human curiosity and the relentless pursuit of knowledge that we continue to explore these tantalizing possibilities, pushing the boundaries of our scientific understanding and challenging our most fundamental assumptions about reality itself. The pursuit of understanding the physics of other realities is not merely a scientific quest; it is a reflection of our deepest human desire to understand our place within the grand cosmic tapestry. The integration of parapsychological observations into this scientific framework requires a cautious and rigorous approach, but the potential insights are transformative. The exploration continues, fueled by both scientific rigor and the boundless human imagination.

8

Consciousness and Other Dimensions

The preceding sections have outlined theoretical frameworks and potential avenues for investigating the existence of other dimensions and realities. However, a truly comprehensive understanding of these possibilities necessitates a deeper exploration of consciousness itself. The nature of consciousness remains one of the most profound and challenging enigmas in science and philosophy. Is consciousness a purely physical phenomenon, a product of the brain's complex neural activity? Or is it something more, something that transcends the physical realm, perhaps even capable of interacting with other dimensions?

The prevailing materialistic view, dominant in contemporary neuroscience, posits that consciousness arises solely from the physical processes within the brain. Neural activity, synaptic connections, and complex patterns of electrical and chemical signaling are considered the sole determinants of subjective experience. While this perspective has yielded significant insights into the neural correlates of consciousness – specific brain regions and activities associated with various conscious states – it leaves unanswered the fundamental question of *how* these physical processes give rise to subjective experience, the feeling of "what it's like" to be conscious. This "hard problem of consciousness," as coined by philosopher David Chalmers, highlights the apparent gap between the objective, physical description of the brain and the subjective, qualitative nature of consciousness.

Integrated Information Theory (IIT), a prominent contemporary theory of consciousness, attempts to address this gap by proposing a measure of consciousness based on the complexity and integration of information within a system. According to IIT, the level of consciousness is directly related to the amount of integrated information a system possesses. Systems with highly integrated information, such as the human brain, are considered to be more conscious than systems with less integrated information. While IIT provides a potentially quantifiable measure of consciousness, it still struggles to fully explain the subjective aspect of experience. It's a step forward, offering a mathematical framework, but it doesn't fully bridge the chasm between physical processes and subjective qualia.

Other theories propose that consciousness is not limited to the brain. Some panpsychist perspectives suggest that consciousness is a fundamental property of the universe, inherent in all matter, albeit at varying degrees of complexity. According to this view, consciousness is not something that "emerges" from complex physical systems, but rather something that is present in all systems, from subatomic particles to galaxies. This perspective offers a radical departure from the materialistic view, suggesting a far more pervasive and fundamental role for consciousness in the cosmos. However, it faces the challenge of explaining how the seemingly disparate aspects of consciousness, from the simple awareness of a single cell to the complex self-awareness of a human being, could all be manifestations of the same fundamental property.

The potential connection between consciousness and other dimensions becomes even more intriguing when we consider the possibility of non-local consciousness. Non-locality, a concept arising from quantum mechanics, suggests that certain quantum phenomena are not limited by spatial distance. Entanglement, for example, implies that two entangled particles can instantaneously affect each other's state, regardless of the distance separating them. Some researchers have speculated that consciousness might also exhibit non-local properties, implying that it might not be confined to the physical brain but could extend beyond the physical realm, perhaps interacting with other dimensions or realities.

Near-death experiences (NDEs) often involve reports of out-of-body experiences, lucid awareness during clinical death, and encounters with deceased loved ones. While skeptics often attribute these experiences to physiological effects of oxygen deprivation or other factors, some researchers suggest they might be indicative of a non-local aspect of consciousness, one capable of persisting beyond the physical limitations of the body. The exploration of NDEs through rigorous scientific investigation, addressing potential biases and alternative explanations, could offer valuable insights into the nature of consciousness and its potential connection to other realms.

Furthermore, the exploration of altered states of consciousness, induced by meditation, psychedelics, or other means, can potentially provide clues to the nature of consciousness and its relationship to other dimensions. These states often involve profound shifts in perception, a sense of interconnectedness with the universe, and experiences that challenge the boundaries of ordinary reality. Again, while caution must be exercised in interpreting these experiences, the possibility of accessing altered states of consciousness that offer glimpses into other dimensions or realities cannot be dismissed. Systematic, controlled studies examining these altered states could yield valuable insights.

The intersection of parapsychology and cosmology provides another potentially fruitful avenue for investigation. Parapsychological phenomena, such as telepathy, clairvoyance, and precognition, challenge conventional understandings of causality and the limitations of physical space and time. If these phenomena are genuine, they could suggest that consciousness is capable of transcending the limitations of the physical body and interacting with information or realities beyond our ordinary perception. A rigorous, scientific investigation of parapsychological phenomena, acknowledging the challenges inherent in researching these controversial topics, could provide invaluable data for understanding the broader nature of consciousness.

The concept of a holographic universe provides yet another fascinating theoretical framework to consider. The holographic principle, initially arising from string theory, suggests that the three-dimensional universe we experience is merely a projection from a two-dimensional surface at its boundary. This concept implies that the universe itself might be a kind of information-processing system, raising profound questions about the nature of reality and the role of consciousness within it. If the universe is holographic, then consciousness might be a fundamental aspect of this information system, inherent in the very structure of reality itself, and potentially capable of accessing other levels or dimensions of that reality.

The relationship between consciousness and other dimensions remains speculative. However, a multidisciplinary approach, combining insights from neuroscience, quantum physics, cosmology, and parapsychology, is essential for unraveling this complex enigma. By exploring the various theoretical frameworks and employing rigorous scientific methods, we can make progress towards a more complete understanding of consciousness, its limits, and its potential connection to realities beyond our ordinary perception. The journey is challenging and necessitates a willingness to consider unconventional ideas while maintaining rigorous scientific standards. Yet, the potential rewards—a deeper understanding of our place in the universe and the very nature of reality—are immense. The exploration continues, driven by a desire to comprehend the ultimate mysteries of existence and our connection to the cosmos as a whole. This involves not only scientific investigation but also a deep philosophical reflection on what consciousness is, and what it means to be aware in a potentially vast and multi-dimensional universe. The convergence of scientific inquiry and philosophical contemplation is vital for this journey, as is the fostering of interdisciplinary collaborations that transcend conventional boundaries and embrace the profound and transformative potential of this research.

Near-death experiences (NDEs) represent a compelling area of inquiry when considering the possibility of consciousness extending beyond the confines of the physical brain. These experiences, often reported by individuals who have clinically died and subsequently been revived, frequently involve a sense of detachment from the body, a feeling of floating or traveling through a tunnel, encounters with deceased loved ones, and profound feelings of peace and transcendence. While some dismiss NDEs as hallucinations caused by oxygen deprivation or other physiological factors associated with near-death situations, the consistency and universality of certain elements within these accounts warrant closer examination.

The subjective nature of NDEs presents a significant challenge for scientific investigation. Unlike objective phenomena that can be readily measured and quantified, NDEs are inherently personal and deeply subjective. The memories and descriptions of these experiences are filtered through the individual's cultural background, personal beliefs, and psychological state. This subjectivity makes it difficult to establish objective criteria for validating or invalidating the claims of those who report them. However, this does not negate the value of investigating these experiences. By employing rigorous methodologies, such as structured interviews, detailed questionnaires, and qualitative analysis, researchers can identify common themes and patterns across a large dataset of NDE accounts. This could offer valuable clues about the underlying mechanisms of NDEs, regardless of whether they are ultimately interpreted as purely physiological or potentially involving a non-physical dimension.

Several hypotheses attempt to explain NDEs within a purely physical framework. One prominent theory suggests that the characteristic features of NDEs, such as the out-of-body experience and vivid imagery, may arise from alterations in brain activity during periods of oxygen deprivation. As the brain's oxygen supply is compromised, neural activity becomes disorganized, leading to hallucinations and perceptual distortions. This hypothesis is supported by some studies demonstrating that certain brain regions, known to be involved in visual processing and spatial awareness, show abnormal activity during near-death situations. However, this explanation does not account for the consistently reported feelings of profound peace, unity, and transcendence, experiences that seem to lie beyond the realm of simple neurological dysfunction.

Other proposed physiological mechanisms involve the release of endorphins, neurochemicals known for their pain-relieving and mood-enhancing effects. The surge of endorphins during periods of stress or trauma might contribute to the feeling of euphoria and detachment often reported in NDEs. Similarly, the release of other neurotransmitters, such as dimethyltryptamine (DMT), a naturally occurring psychedelic compound found in the human brain, has been suggested as a possible contributor to the hallucinatory aspects of NDEs. These physiological explanations, while offering plausible mechanisms for some aspects of NDEs, fail to fully account for the highly consistent and often life-changing nature of these experiences.

In contrast to purely physiological interpretations, some researchers suggest that NDEs may provide evidence for the existence of consciousness independent of the brain. This view aligns with certain non-materialist theories of consciousness, which propose that consciousness is not merely a product of brain activity but a fundamental aspect of reality. From this perspective, NDEs might represent a temporary detachment of consciousness from the physical body, allowing it to experience realms or dimensions beyond the limitations of the physical world. This interpretation aligns with the idea of a non-local consciousness, capable of existing independently of space and time. However, such a perspective remains highly speculative and requires further investigation to validate its plausibility.

The exploration of NDEs necessitates a multidisciplinary approach, combining insights from neuroscience, psychology, and potentially even parapsychology. Neuroscientific research can help identify the neural correlates of NDEs, while psychological research can explore the individual factors that shape the subjective experiences. Furthermore, parapsychological research, while often controversial, might offer valuable insights into the potential for consciousness to interact with non-physical realities. By integrating these diverse perspectives, we can approach NDEs with a greater degree of scientific rigor, while maintaining an open mind to the possibility of experiences that transcend our current scientific understanding.

Beyond NDEs, other altered states of consciousness can provide further insights into the potential connection between consciousness and other dimensions. Experiences induced by meditation, deep hypnosis, sensory deprivation, and the use of psychedelics often involve profound alterations in perception, heightened sensory awareness, and a sense of unity or interconnectedness with the universe. These experiences often challenge the boundaries of ordinary reality, leading some to speculate that they might represent glimpses into other dimensions or states of being. However, it is important to approach these experiences with caution, recognizing the potential for subjective bias, misinterpretations, and the influence of cultural and psychological factors.

The potential for consciousness to interact with other dimensions or realities raises profound implications for our understanding of the universe. It challenges the conventional materialistic view, which posits that the physical world is the sole reality and that consciousness is merely a by-product of physical processes. If consciousness can indeed exist independently of the brain or extend beyond the physical realm, it opens up the possibility of a far more expansive and interconnected reality than we currently comprehend. This necessitates a shift in our scientific paradigms, moving beyond the limitations of reductionist approaches to embrace a more holistic and integrated view of the universe. This requires not only rigorous scientific investigation, but also a willingness to consider alternative explanations, challenge established assumptions, and explore uncharted territory.

The journey toward understanding the potential connection between consciousness and other dimensions is challenging but potentially transformative. It requires a multidisciplinary approach, bringing together insights from neuroscience, physics, cosmology, psychology, and philosophy. By embracing rigorous scientific methodology, while maintaining an open mind to the possibility of extraordinary phenomena, we can incrementally advance our understanding of this profound mystery. The potential rewards are immense, potentially leading to a profound shift in our understanding of the nature of reality, the place of consciousness within it, and our connection to the broader cosmos. The implications are far-reaching, impacting not only our understanding of science but also our philosophical perspectives, ethical considerations, and spiritual beliefs.

Out-of-body experiences (OBEs), often described as a sensation of existing outside one's physical body, present a fascinating area of exploration when considering the possibility of interdimensional travel or interaction with other realities. While the neurological underpinnings of OBEs are still under investigation, the consistent reporting of certain features across diverse cultures and individuals warrants further scrutiny beyond purely physiological explanations. Many individuals report a sense of floating above their body, observing their surroundings from a detached perspective, often with an enhanced sense of spatial awareness. Some accounts describe the ability to travel to distant locations, or even to other perceived planes of existence, during these experiences.

One common theory links OBEs to disruptions in the brain's processing of sensory information. Disruptions in the temporoparietal junction (TPJ), a brain region crucial for integrating sensory input and maintaining a sense of self, have been implicated in OBEs. Brain imaging studies during OBEs, although limited, suggest abnormal activity in this area. However, this does not fully explain the reported experiences of travelling beyond the immediate physical environment, suggesting the possibility of mechanisms beyond simple sensory misinterpretations. Furthermore, the feeling of heightened awareness and clarity often associated with OBEs contradicts the expected disorientation and confusion associated with neurological disturbances.

The potential role of altered states of consciousness in OBEs is also relevant. Many individuals report experiencing OBEs during periods of sleep paralysis, a condition where individuals are temporarily unable to move or speak while either falling asleep or waking up. This state is characterized by a heightened sense of awareness, coupled with vivid hallucinations and a feeling of being trapped. Sleep paralysis often involves sensory distortions and a heightened sense of fear, although OBEs experienced during sleep paralysis are often described as more peaceful and insightful. This suggests that the underlying mechanisms triggering OBEs may be distinct from those involved in sleep paralysis, despite their frequent co-occurrence.

Another intriguing aspect of OBEs is the frequent reporting of lucid dreaming, a state of consciousness where individuals are aware they are dreaming and can actively influence the dream's narrative. Many individuals report transitioning from lucid dreams into OBEs, blurring the line between the dream state and the waking state. This suggests that the mechanisms underpinning OBEs may share similarities with those responsible for lucid dreaming, potentially involving a decoupling of consciousness from the physical body and the activation of other cognitive processes typically suppressed during normal waking consciousness.

The possibility of OBEs representing glimpses into other dimensions or realities is a topic of significant speculation. Some parapsychologists suggest that OBEs might involve a temporary separation of consciousness from the physical body, allowing individuals to access other planes of existence, or even to traverse different levels of reality. This aligns with certain theories in quantum physics, suggesting that our reality might be composed of multiple overlapping dimensions or levels of reality, potentially accessible through altered states of consciousness. This perspective requires a shift from a purely materialistic worldview, accepting the possibility of non-physical dimensions and the capacity of consciousness to interact with them.

The concept of interdimensional travel, while seemingly fantastical, is not entirely outside the realm of theoretical physics. String theory, for example, postulates the existence of multiple spatial dimensions beyond the three we experience in everyday life. These extra dimensions are thought to be curled up and compactified, making them imperceptible to our senses. However, under specific conditions, such as those potentially encountered during altered states of consciousness, these dimensions might become accessible, leading to experiences like OBEs that transcend the limitations of our ordinary perception of reality.

The exploration of OBEs requires a multidisciplinary approach, encompassing neuroscience, psychology, and potentially even physics and parapsychology. Neuroscientific studies can investigate the neural correlates of OBEs, identifying the brain regions and neurochemical processes involved. Psychological research can explore the cognitive and emotional factors shaping the experience, examining the role of expectation, belief, and individual personality traits in shaping the subjective experience.

Parapsychological research, while often controversial within mainstream science, can provide valuable insights into the potential connection between OBEs and other paranormal phenomena, such as telepathy, clairvoyance, and precognition. These phenomena, if genuinely existent, might suggest the possibility of non-local consciousness, capable of extending beyond the constraints of space and time, potentially providing a mechanism for interdimensional travel or interaction with other realities. However, critical scrutiny and rigorous methodology are paramount to avoid spurious correlations and subjective interpretations.

However, rigorous scientific validation of claims related to interdimensional travel remains elusive. The subjective nature of OBEs makes objective verification challenging. The experiences are highly personal and influenced by individual beliefs and cultural background, making it difficult to establish universally accepted criteria for authentication. Furthermore, the lack of independent verification and the difficulty in replicating OBEs under controlled laboratory conditions pose significant hurdles for rigorous scientific investigation.

Despite these challenges, exploring OBEs and their potential connection to interdimensional travel serves as a catalyst for broader scientific inquiry. It compels us to question our understanding of consciousness, the nature of reality, and the potential limits of human perception. By integrating insights from various disciplines, employing rigorous research methodologies, and maintaining an open yet critical approach, we can gradually unravel the mysteries surrounding these profound experiences, moving beyond simplistic explanations and acknowledging the possibility of phenomena that extend beyond our current scientific understanding.

Furthermore, the investigation into OBEs and their potential link to interdimensional travel could have significant implications for our understanding of cosmology and the multiverse. If consciousness is not limited to the physical body, as some OBEs might suggest, it could potentially exist independently of space and time, transcending the boundaries of our universe and even interacting with other universes within a multiverse framework. This has profound implications for our conception of life, death, and the nature of existence itself.

The ongoing exploration of OBEs requires a balance between scientific rigor and open-mindedness. Dismissing the experiences out of hand due to a lack of current scientific explanations is as limiting as accepting all reported accounts without critical evaluation. A cautious, yet curious, approach is needed, utilizing rigorous scientific methodology to investigate these experiences while remaining open to the possibility of phenomena that currently defy our understanding. The journey of exploring the mysteries of OBEs and interdimensional travel is a journey into the unknown, fraught with challenges, yet rich in potential for uncovering profound truths about the nature of reality and our place within it. The implications of discovering a credible link between OBEs and other dimensions are potentially transformative, reshaping our understanding of consciousness, the universe, and our place within the cosmos.

The exploration of out-of-body experiences and their potential connection to other dimensions naturally leads us to consider the intriguing interplay between consciousness and quantum physics. The very nature of consciousness remains one of the greatest unsolved mysteries in science, and its connection to the quantum realm offers a particularly fertile ground for speculation and research. While a definitive link remains elusive, the parallels between certain aspects of quantum mechanics and subjective experiences are compelling enough to warrant serious consideration.

One of the most striking parallels lies in the concept of superposition, a fundamental principle in quantum mechanics. Superposition dictates that a quantum system, such as an electron, can exist in multiple states simultaneously until measured. This means the electron isn't definitively in one location or another until it's observed; the act of observation seems to "collapse" the wave function, forcing the electron to adopt a specific state. This has led some physicists and philosophers to suggest that consciousness plays a crucial role in the collapse of the wave function, implying that observation, as a conscious act, is an integral part of the quantum measurement process.

This idea is further supported by the concept of quantum entanglement, where two or more quantum particles become linked in such a way that they share the same fate, regardless of the distance separating them. Measuring the state of one entangled particle instantly determines the state of the other, even if they are light-years apart. This "spooky action at a distance," as Einstein famously called it, suggests a non-local connection between entangled particles, transcending the limitations of space and time. Some theorists suggest that consciousness might be similarly non-local, capable of interacting with the quantum world in ways that defy our classical understanding of causality and locality.

The implications of a conscious observer influencing quantum events are profound. It challenges the traditional view of the universe as a deterministic system governed by fixed laws of physics. If consciousness can affect quantum measurements, it implies a degree of agency and free will that isn't readily accounted for in classical physics. It suggests that our subjective experiences might actively shape the reality we perceive, rather than passively observing a pre-determined universe.

However, the concept of consciousness influencing quantum events faces significant challenges. The exact mechanism by which consciousness might interact with quantum systems remains completely unknown. Many physicists argue that the collapse of the wave function is a consequence of the interaction between the quantum system and its environment, not a conscious act of observation. The measurement problem in quantum mechanics is still a hotly debated topic, and the role, if any, of consciousness remains far from settled.

Furthermore, even if consciousness does influence quantum events, the scale of this influence remains a matter of speculation. Quantum effects are typically observed at the microscopic level, involving individual particles or small systems. It's unclear how these effects could scale up to influence macroscopic objects or even human consciousness. The transition from quantum phenomena to classical behaviour, known as decoherence, occurs extremely rapidly in macroscopic systems, making it difficult to envision how consciousness could maintain coherence and interact with the quantum realm in a meaningful way.

Despite these challenges, the exploration of consciousness and quantum physics continues to fascinate scientists and philosophers alike. Various theories have been proposed to explain the potential link between consciousness and the quantum realm, including those that posit a fundamental role for consciousness in the structure of reality itself. These theories often involve extending or modifying existing quantum theories to accommodate the role of consciousness, suggesting that consciousness may be a fundamental aspect of the universe, rather than an emergent property of complex physical systems.

One such theory is Orchestrated Objective Reduction (Orch-OR), proposed by Roger Penrose and Stuart Hameroff. This theory suggests that consciousness arises from quantum computations occurring within microtubules, structures found inside neurons. According to Orch-OR, these quantum computations are orchestrated by the structure of the microtubules and lead to objective reductions of the wave function, resulting in conscious experience. While Orch-OR has garnered considerable attention and debate, it has also faced criticism due to the challenges in demonstrating the existence of quantum coherence in microtubules and the mechanisms by which these quantum processes could give rise to consciousness.

Another line of inquiry involves the exploration of quantum entanglement and its potential relevance to non-local consciousness. Some researchers have suggested that entanglement could provide a mechanism for non-local correlations between conscious minds, potentially explaining phenomena like telepathy or shared consciousness. However, replicating and verifying these phenomena under controlled experimental conditions has proven incredibly difficult, and alternative explanations, such as subtle sensory cues or biases in data interpretation, cannot be easily ruled out.

The study of near-death experiences (NDEs) also offers intriguing potential insights into the consciousness-quantum physics connection. Many individuals who have experienced NDEs report feelings of detachment from their physical body, a sense of profound peace, and vivid perceptions of their surroundings, despite their physiological state. Some have even reported out-of-body experiences, similar to OBEs but often characterized by a feeling of increased consciousness and awareness. These experiences have led some researchers to speculate that consciousness might persist beyond the confines of the physical body, perhaps interacting with the quantum realm in ways that defy our current understanding.

However, interpreting NDEs within a scientific framework faces significant challenges. NDEs are subjective and personal experiences, highly influenced by individual beliefs and cultural context. The lack of objective verification and the potential influence of physiological factors, such as oxygen deprivation, complicate the search for rigorous scientific explanations. It is crucial to differentiate between verifiable physical changes and subjective perceptions reported by individuals undergoing profound physiological shifts.

Despite these challenges, the exploration of consciousness and quantum physics remains a highly active area of research and speculation. Advances in quantum computing, neuroscience, and our understanding of the mind-brain connection could potentially offer new insights into this fascinating interplay. While a definitive answer remains elusive, the potential implications of a fundamental link between consciousness and the quantum realm are immense, potentially revolutionizing our understanding of reality, the nature of consciousness, and our place in the universe. The journey continues, blending the rigorously scientific with the boldly speculative, seeking to unlock the profound mysteries at the intersection of the subjective and the quantum. This requires an interdisciplinary approach, bridging the gap between physics, neuroscience, philosophy, and even parapsychology, with careful consideration of both experimental data and theoretical possibilities.

The study of altered states of consciousness, beyond OBEs and NDEs, also presents an avenue for exploring the link between mind and quantum phenomena. Meditation, hypnosis, and psychedelic experiences all involve changes in brain activity and perception that could potentially influence the interaction between consciousness and the quantum realm. While the research in these areas is still in its infancy, preliminary findings suggest that altered states of consciousness may be associated with changes in brainwave patterns and neurochemical activity that could be relevant to the quantum measurement problem. These changes might involve a modulation of neuronal firing patterns that influences quantum coherence within the brain, or possibly changes in the brain's electromagnetic fields that could interact with quantum systems. Further investigation is needed to understand the potential connection between altered states of consciousness and the quantum world.

Ultimately, the relationship between consciousness and quantum physics remains a mystery, but one that is increasingly attracting the attention of scientists and philosophers alike. The potential implications of finding a definitive link are staggering, potentially transforming our understanding of the fundamental nature of reality and our place within it. A balanced approach is necessary, integrating rigorous scientific methodology with open-mindedness and a willingness to explore unconventional theories. The journey to unlock these profound secrets promises to be a long and challenging one, but the potential rewards are immense. The quest to understand the intersection of consciousness and quantum physics is a quest to understand the very fabric of reality itself, a journey towards a deeper and more complete understanding of our existence.

The exploration of consciousness's potential role in shaping reality necessitates a deep dive into philosophical implications that extend far beyond the realm of physics. The very notion that our subjective experiences might not merely reflect an objective reality, but actively participate in its construction, challenges deeply ingrained assumptions about the nature of existence. This challenges the classical mechanistic worldview, which posits a universe operating solely according to predetermined laws, independent of any observer. If consciousness plays an active role in shaping reality, then the universe becomes a far more dynamic and participatory system.

One of the key philosophical considerations is the concept of idealism versus realism. Realism traditionally holds that an external reality exists independently of our perception of it. Idealism, conversely, suggests that reality is fundamentally mental, a construct of consciousness. While extreme idealism might propose that reality is entirely a product of individual minds, a more nuanced perspective suggests that consciousness, as a collective phenomenon, plays a fundamental role in shaping the shared reality we experience. The implications of this are vast, potentially redefining our understanding of objectivity and truth. If reality is partially constructed by consciousness, then what we perceive as "objective" might be, to some degree, a product of collective agreement or shared perception.

This participatory aspect of reality brings forth questions regarding free will and determinism. The classical deterministic view suggests that all events are predetermined by prior causes, leaving little room for conscious agency. However, if consciousness actively participates in shaping reality, it suggests a degree of freedom and influence on the course of events. This doesn't necessarily imply total freedom from causal influence; rather, it suggests that consciousness might act as a catalyst, influencing the probabilities of various outcomes within a framework of underlying laws. This perspective aligns with many quantum interpretations which view the universe as probabilistic rather than deterministic.

Furthermore, the role of consciousness in shaping reality raises questions about the nature of time. Our subjective experience of time is linear and sequential, a continuous flow from past to future. However, some interpretations of quantum mechanics, and certain mystical traditions, suggest that time might be less rigid than we perceive it. If consciousness plays a role in constructing reality, it might be that our experience of linear time is merely one interpretation, a projection of our own consciousness onto the underlying fabric of reality. This leads to the consideration of whether our conscious choices might influence not only the present, but also the past and the future, challenging the linear narrative of time.

The concept of a multiverse, often discussed in cosmology, further complicates our understanding of consciousness's role in shaping reality. If multiple universes exist, each with its own unique physical laws and configurations, how might consciousness relate to these different realms? Does consciousness exist independently of any particular universe, potentially traversing or influencing multiple realities? These are questions that push the boundaries of both physics and philosophy, requiring us to expand our conceptual frameworks to accommodate the immense possibilities implied by the existence of a multiverse.

The potential connection between consciousness and the quantum realm also presents complex philosophical challenges. The uncertainty principle in quantum mechanics suggests that certain properties of quantum systems, such as position and momentum, cannot be simultaneously known with perfect accuracy. This uncertainty is not merely a limitation of our measurement techniques, but a fundamental aspect of the quantum world. Some theorists suggest that consciousness plays a role in "collapsing" the wave function, choosing a particular outcome from a range of possibilities. This introduces a subjective element into what is typically viewed as an objective scientific realm, raising profound questions about the nature of objectivity and the relationship between the observer and the observed.

The concept of quantum entanglement, where two or more quantum particles become linked in such a way that their fates are intertwined, even across vast distances, presents further philosophical challenges. Some interpretations suggest that entangled particles might be connected through a non-local consciousness field, transcending the limitations of space and time. This concept challenges our classical understanding of causality and locality, raising questions about the fundamental nature of space, time, and consciousness itself. It might imply the existence of a deeper level of reality, beyond our everyday perception, where consciousness plays a fundamental role in the interconnectedness of all things.

The philosophical implications of consciousness's role in shaping reality are profound and far-reaching. They necessitate a reassessment of our fundamental assumptions about the nature of existence, challenging our classical mechanistic worldview. They require us to consider the interconnectedness of all things, the potential for non-locality, and the complex relationship between the observer and the observed. While conclusive answers remain elusive, the exploration of these philosophical questions is crucial for a deeper understanding of reality and our place within it. It demands an interdisciplinary approach, combining insights from physics, philosophy, and even areas like parapsychology, to tackle the profound mysteries at the intersection of consciousness and the cosmos. This necessitates a shift from merely passive observation to active participation in the ongoing process of creating and shaping our reality. The journey is one of intellectual exploration, challenging our preconceived notions and pushing the boundaries of human understanding. It is a quest to uncover the deep mysteries at the heart of existence itself, a journey that unites the scientific rigor with the boundless imagination.

The search for answers continues, driving us to delve deeper into the enigmatic interplay between consciousness and the universe, ultimately transforming our understanding of ourselves and our place within the grand tapestry of existence. This involves acknowledging the limitations of our current understanding, embracing uncertainty, and maintaining an open mind as we navigate the uncharted territories of consciousness's potential influence on reality. It's a path of constant questioning, exploration, and reassessment, a journey of intellectual and spiritual discovery that leads us closer to a more complete and profound understanding of the universe and our intricate relationship with it.

9

Anomalies and Unexplained Phenomena

The preceding discussion of consciousness and its potential influence on reality provides a fertile ground for examining unexplained phenomena. While mainstream science often dismisses these anomalies, a framework that considers the possibility of multiple densities, realms, and dimensions offers alternative perspectives that warrant investigation. By applying the principles outlined earlier—the interplay of consciousness, density variations, and potential interdimensional interactions—we can approach these enigmas with a renewed sense of scientific curiosity rather than outright dismissal.

One such anomaly is spontaneous human combustion (SHC), a phenomenon where a person allegedly burns to ashes without any apparent external source of ignition. The sheer improbability of this event, given our understanding of human anatomy and the physics of combustion, has led to its classification as a fringe topic. However, a multidimensional perspective might offer a plausible explanation. Consider the possibility of a localized density fluctuation, a temporary breach in the fabric of reality, where the individual's body interacts with a higher-energy realm. This interaction could manifest as an instantaneous and intense combustion, leaving behind minimal physical evidence. The purported lack of surrounding damage in some documented cases of SHC could be explained by the highly localized nature of such a density fluctuation—a "rippling" effect contained within a specific spatial boundary, leaving the surrounding environment unaffected.

While documented cases of SHC are often scant on detail and subject to varying interpretations, the consistency of certain features—rapid and complete incineration, minimal external damage—suggests the need for a deeper, more holistic analysis. The role of consciousness in this process should also be considered. Could intense emotional states, a profound shift in the individual's consciousness, act as a catalyst for such a dimensional interaction? This is, admittedly, highly speculative, but it opens avenues for exploration that traditional scientific explanations might overlook.

Crop circles, another set of unexplained phenomena, defy easy categorization. The intricate geometric patterns appearing in fields overnight, often with precision impossible to replicate using conventional tools, challenge our understanding of natural formations. Some researchers propose mundane explanations, suggesting the patterns are human-made hoaxes. However, many documented cases exhibit complexities that remain unexplained even with such explanations. The precision, scale, and speed of creation suggest a technology or process currently beyond our comprehension. Could this be evidence of an advanced technology, perhaps not from this density or dimension? Could conscious intention, either human or otherwise, be playing a role in their creation, acting as a form of interdimensional communication?

Furthermore, the prevalence of crop circles around ley lines, areas of supposed high-energy convergence, may support the hypothesis of interdimensional interaction. These areas could be points of dimensional instability, where interactions between different densities are more readily apparent. The suggestion of an energetic correlation between crop circles and ley lines demands further scientific investigation. While currently lacking conclusive evidence, these hypotheses open up new avenues of research. Precise geospatial mapping of crop circle occurrences, coupled with thorough analysis of energy fields around these locations, might reveal correlations previously overlooked. The integration of advanced imaging technology and statistical analysis offers the potential to provide quantitative data to strengthen the evidence.

Beyond crop circles and spontaneous human combustion, many other anomalies demand attention. Reports of poltergeist activity, unexplained disappearances, and seemingly inexplicable technological malfunctions raise similar questions. These seemingly disparate phenomena might share common threads, united by their apparent defiance of conventional scientific understanding. The recurring theme suggests a possibility that the underlying laws governing these anomalies are not yet fully understood.

One potential explanation lies in the concept of interdimensional bleed-through—the intrusion of phenomena from other densities or dimensions into our own. This bleed-through could manifest in different ways, depending on the nature of the interacting densities and the level of the interaction. Small-scale interactions might cause seemingly random events, while larger-scale interactions could manifest as more significant anomalies. The concept of a multidimensional universe, with different densities and realms interacting in various ways, provides a theoretical framework for explaining events seemingly outside of conventional physics. The key lies in recognizing and acknowledging the limitations of our current scientific paradigms. The exploration of anomalies requires an openness to alternative hypotheses, a willingness to embrace the unknown and to challenge existing dogmas.

To further analyze these events, a multidisciplinary approach is crucial. The integration of expertise from fields such as physics, parapsychology, anthropology, and computer science is essential to create a comprehensive understanding. Detailed and rigorous documentation of anomalous events, employing standardized data-collection methods and advanced analytical tools, is crucial. This approach, focusing on evidence-based investigation, will be more likely to yield objective findings and help to distinguish between plausible and improbable explanations.

The process of systematically analyzing unexplained events must also incorporate a rigorous methodology, akin to the scientific method applied to other areas of research. This involves formulating testable hypotheses, designing experiments to test these hypotheses, and rigorously analyzing the data obtained. The focus should be on replicating findings, thereby ensuring the reliability and validity of any conclusions drawn. The exploration of anomalies demands a critical and skeptical mindset; extraordinary claims require extraordinary evidence. However, such skepticism should not preclude open-mindedness, the willingness to consider alternative hypotheses.

The exploration of these anomalies is not merely an exercise in intellectual curiosity; it represents a potential paradigm shift in our understanding of the universe and our place within it. By acknowledging the limitations of our current scientific models and embracing the possibility of alternative realities, we can move towards a more comprehensive understanding of the cosmos. This understanding may eventually reveal that the anomalies are not aberrations but rather manifestations of a reality far richer, more complex, and far stranger than we ever imagined. The implications of such a discovery would be profound, transforming not only our scientific knowledge but also our philosophical and spiritual perspectives. This journey requires a commitment to rigorous investigation and critical analysis, balanced by an openness to possibilities that currently lie beyond the boundaries of our established scientific understanding. The willingness to engage with these seemingly inexplicable phenomena, to question assumptions, and to explore alternative perspectives is crucial for unlocking the secrets of the universe and enhancing our comprehension of the enigmatic interconnectivity of all things.

The exploration of anomalies naturally leads us to consider the profound influence of the mind, specifically the power of belief, on physical reality. This is perhaps most clearly illustrated by the placebo effect, a phenomenon where a seemingly inert substance or procedure produces a therapeutic effect simply because the recipient believes it will. While often dismissed as a purely psychological effect, the placebo effect challenges our simplistic understanding of cause and effect, suggesting a deeper connection between mind and body, and potentially, mind and reality itself.

Numerous studies have demonstrated the remarkable power of the placebo effect. In clinical trials, placebo treatments have shown significant efficacy in alleviating symptoms ranging from pain and anxiety to depression and even Parkinson's disease. The magnitude of these effects can be surprisingly substantial, sometimes rivaling or even exceeding the effects of active medications. For example, studies have shown that placebo analgesics can reduce pain perception as effectively as morphine in some patients. This is not simply a matter of suggestion; brain imaging studies have revealed that placebos can activate the same neural pathways as active drugs, indicating a genuine physiological response.

The mechanisms underlying the placebo effect remain a subject of ongoing research. However, several factors appear to be crucial. One is expectation; the stronger the belief in the efficacy of the treatment, the more pronounced the effect. This suggests a crucial role for conscious intention and the power of belief. Furthermore, the context in which the placebo is administered plays a significant role. A physician's demeanor, the setting of the treatment, and the overall patient experience can all influence the efficacy of the placebo. This highlights the importance of the mind-body connection and the influence of external factors on the internal physiological response.

The placebo effect is not limited to medical contexts. It's been observed in a wide range of situations, suggesting a fundamental principle at play. For example, studies have shown that placebo interventions can improve athletic performance, enhance cognitive function, and even influence immune responses. These findings suggest that the placebo effect is not a mere quirk of the medical system, but rather a fundamental aspect of the interaction between mind and body.

But the implications of the placebo effect extend far beyond the realm of medicine. It raises profound questions about the nature of reality itself. If a belief can produce a measurable physical effect, as demonstrated by the placebo effect, then it suggests that our conscious minds might be capable of influencing reality in ways we don't yet fully understand. This notion challenges the traditional materialistic view of the universe, which posits that matter is primary and consciousness is merely an epiphenomenon—a secondary effect of brain activity. The placebo effect suggests a more interactive model, where consciousness can actively participate in shaping physical reality.

This interaction could be explained, within the framework of multidimensional physics, by considering the nature of consciousness itself. As previously discussed, consciousness may not be limited to the physical brain; it may exist in a higher dimensional space, influencing the physical realm through subtle energetic interactions. The belief in a treatment could then be viewed as a form of conscious intention, modulating the flow of energy within the multidimensional field. This modulation could then produce measurable effects in the physical world, manifested as the placebo effect.

Furthermore, the placebo effect aligns with other phenomena that suggest a connection between consciousness and reality. Examples include remote viewing, where individuals accurately perceive distant locations, and psychokinesis, where the mind seems to influence physical objects. While these phenomena are highly controversial and require further investigation, they align with the idea that consciousness is not merely a passive observer of reality but an active participant. These instances highlight a complex interplay between observer and observed, challenging the traditional separation between subject and object. The consistency across different observed phenomena highlights a potential underlying mechanism.

To deepen our understanding of the placebo effect and its implications, a more comprehensive, interdisciplinary approach is needed. It requires collaboration between physicists, neuroscientists, psychologists, and other researchers to explore the underlying mechanisms, using advanced imaging techniques and sophisticated data analysis. This research should not only focus on the biological and psychological aspects of the placebo effect, but also explore its potential implications within a broader framework of consciousness studies. This might include exploring the potential influence of collective consciousness and the role of subtle energies in shaping physical reality.

One promising avenue of research involves exploring the role of quantum mechanics in the placebo effect. Quantum mechanics deals with the interaction between consciousness and matter, the observation of reality affecting the measured state. Could the act of belief, a conscious act, trigger quantum effects within the body, leading to the observed therapeutic effects? The potential for quantum entanglement between the observer and the observed offers a unique perspective within this framework, with the observer's belief impacting the observed physical response. This requires a deeper understanding of the quantum processes within the human body and the potential interactions with the surrounding environment.

Furthermore, the study of subtle energy fields, like those associated with acupuncture meridians, warrants further exploration. Could placebos work by influencing these subtle energies, thereby affecting the body's physiology? The integration of traditional healing modalities with modern scientific methods may yield new insights into the mechanisms of the placebo effect and its wider implications. Research across disciplines is necessary to bridge the gap between these seemingly disparate realms of knowledge.

The power of belief, as demonstrated by the placebo effect, has profound implications not only for medicine but also for our understanding of consciousness and reality. It challenges the materialistic worldview and opens up possibilities for a more holistic view of the universe, one where mind and matter are intricately interwoven. This calls for a paradigm shift in our scientific understanding, encouraging a more integrative approach to research that embraces both conventional science and alternative perspectives. The exploration of the placebo effect is not just a quest to understand a peculiar medical phenomenon; it's a journey into the very nature of reality itself, a path that may ultimately lead us to a deeper understanding of our place within the cosmos and the extraordinary power of our minds. By continuing to investigate these phenomena with rigorous scientific methodology, while maintaining an open mind to the possibilities that currently lie outside of the boundaries of our established scientific understanding, we may unlock profound truths about the very fabric of existence. The implications of such a discovery would be transformative, extending far beyond the realm of science to reshape our philosophical and spiritual understandings of the universe.

The exploration of the placebo effect and its implications for our understanding of consciousness and reality naturally leads us to consider a broader range of unexplained phenomena. While the placebo effect demonstrates a mind-body connection capable of influencing physical reality within our perceived dimension, other anomalies suggest potential interactions with realms beyond our immediate sensory experience. These interactions, often relegated to the realm of speculation and parapsychology, warrant a more rigorous scientific examination, particularly within the framework of multidimensional physics and cosmology.

One such area of investigation involves the numerous documented cases of spontaneous human combustion (SHC). While often dismissed as myths or misinterpretations of other events, the persistent reports of individuals seemingly combusting without an external ignition source present a challenge to our current scientific understanding. The lack of consistent physical evidence, such as the presence of large-scale fire damage surrounding the victim, further fuels the mystery. However, could some phenomena such as SHC be explained through the lens of interdimensional interactions?

Perhaps, under specific, yet currently unknown conditions, an individual's energy field becomes unstable, allowing for a catastrophic interaction with a higher-dimensional energy source. This interaction, rather than causing a conventional fire, could directly convert the individual's matter into energy, leaving limited physical trace. This hypothesis requires a deeper understanding of higher-dimensional energy fields and their potential interactions with matter in our four-dimensional spacetime. It would also require new scientific tools to measure and analyze these phenomena safely and effectively.

Similarly, poltergeist activity, characterized by unexplained movements of objects, sounds, and other disturbances, presents another intriguing avenue of exploration. While many instances are attributed to psychological factors or mundane explanations, other cases defy easy categorization. Could these phenomena be manifestations of interdimensional energies or entities interacting with our reality? The seemingly intelligent nature of some poltergeist activities, including the targeted manipulation of specific objects or the creation of seemingly purposeful messages, suggests a level of coordination that surpasses simple random events. This requires investigating the potential for intelligent life forms existing in higher dimensional spaces, interacting with our physical world in unpredictable ways. Exploring the nature of these interactions might provide valuable insights into how energy and matter interact across multiple dimensions.

The investigation into these phenomena requires a paradigm shift from traditional reductionist approaches, moving toward a more holistic, systems-based perspective. This paradigm shift requires us to revisit fundamental assumptions about the nature of space, time, energy, and matter. Existing physics may prove incomplete to explain these apparent anomalies. The possibility of multidimensional realities, as explored in string theory and M-theory, necessitates a reassessment of how we perceive and interact with our environment. For example, if our reality is but one 'brane' within a multidimensional multiverse, interactions between these branes could cause the appearance of anomalous events in our spacetime. These interactions might involve an exchange of energy or information, manifesting as unexplained phenomena in our physical world.

Furthermore, near-death experiences (NDEs) provide another compelling area of inquiry. The consistent reports of individuals experiencing out-of-body sensations, encounters with deceased loved ones, and vivid descriptions of other realms during near-death episodes warrant careful consideration. While neurological explanations exist to account for some aspects of NDEs, the universality of certain themes and the rich descriptive detail provided by individuals who have undergone such experiences suggest the possibility of consciousness transcending the physical limitations of the brain. These experiences could represent a glimpse into other dimensions or realities, revealing a greater continuity of consciousness beyond physical death. We might need to develop novel instruments and experimental paradigms to approach this exploration in a responsible and scientific manner. This entails a transdisciplinary approach involving neuroscientists, physicists, and parapsychologists.

The intersection of cosmology and parapsychology presents a potentially fruitful avenue of research. The vastness of the universe, the possibility of other universes, and the presence of dark matter and dark energy, all suggest that our current understanding of the cosmos is far from complete. Many cosmological models predict the existence of multiple dimensions, and the interaction of these dimensions with our own could manifest as unexplained phenomena. Moreover, advanced civilizations inhabiting other dimensions might possess technologies or abilities beyond our current comprehension, potentially influencing our reality in ways that we currently fail to detect. Further analysis of unexplained occurrences might reveal underlying patterns or correlations that hint at interdimensional influences.

The study of unexplained events requires a multidisciplinary approach. Physicists can contribute by developing theoretical frameworks to explain these occurrences, while neuroscientists can explore the neurological mechanisms that might underlie some of these experiences. Parapsychologists, while sometimes marginalized within the scientific community, can offer valuable insights based on their long-term study of these phenomena. A collaborative effort, integrating insights from various disciplines, is crucial to advancing our understanding of these anomalies.

Moreover, the ethical implications of interdimensional interactions require careful consideration. If we can demonstrate that unexplained events are indeed the result of interactions with other realities or dimensions, then the potential consequences of such interactions must be fully assessed. We need to establish guidelines and protocols to ensure responsible exploration of these phenomena, avoiding potential dangers or unintended consequences. Developing a framework that allows responsible investigation while minimizing risk is crucial for moving forward ethically.

In conclusion, while unexplained events remain enigmatic, they present a significant challenge to our current scientific paradigm. By adopting a more open-minded and interdisciplinary approach, embracing the possibilities of multidimensional physics and cosmology, we can begin to unravel the mysteries surrounding these phenomena. This requires a willingness to challenge traditional assumptions, to explore new avenues of research, and to collaborate across different scientific disciplines. The exploration of unexplained events is not simply a matter of satisfying intellectual curiosity; it is a journey toward a deeper understanding of reality, of our place within the cosmos, and of the vast potential of consciousness itself. This interdisciplinary and rigorously scientific approach could lead to significant breakthroughs, transforming our understanding of the universe and our place within it. The journey, however challenging, holds the promise of incredible discoveries and fundamentally shifts our understanding of the cosmos and the nature of reality.

The previous discussion highlighted the existence of anomalous phenomena that challenge our conventional understanding of reality. To move beyond mere speculation, we must establish rigorous methodologies for their investigation. This necessitates a critical approach, minimizing subjective biases and maximizing objectivity. The challenge lies in applying the scientific method to phenomena that, by their very nature, resist easy categorization and often elude conventional instruments.

One crucial aspect is the rigorous documentation of anomalous events. This involves detailed accounts of the circumstances surrounding the event, including time, location, witnesses, and any physical evidence. High-resolution photography, videography, and audio recording, where applicable, are essential. However, the reliance on technology alone is insufficient; meticulous observation and detailed written descriptions remain paramount, particularly in instances where technology is unavailable or unreliable.

The issue of confirmation bias poses a significant hurdle. Investigators must actively strive to avoid interpreting data in a way that supports pre-existing beliefs. This requires a conscious effort to consider alternative explanations, however improbable they may seem initially. A robust investigation involves a systematic exploration of all potential explanations, including mundane possibilities, before resorting to more extraordinary hypotheses. This process often necessitates employing control groups, statistical analysis, and blind experiments to mitigate the influence of subjective biases. In cases involving subjective experiences, such as near-death experiences or remote viewing, the incorporation of standardized questionnaires and psychological assessments is vital in identifying potential confounding factors.

The development of new instrumentation for detecting and measuring anomalous phenomena is another critical area. Many reported events lack readily available methods of measurement. For example, the elusive nature of some poltergeist phenomena necessitates the development of highly sensitive sensors to detect subtle changes in electromagnetic fields, temperature, or acoustic pressure. Similarly, the investigation of spontaneous human combustion requires advances in forensic science, possibly involving new techniques for analyzing trace evidence and residual energy patterns. Technological advancements are crucial in transitioning the study of anomalous phenomena from anecdotal observation to quantitative analysis.

A multidisciplinary approach is paramount. The investigation of unexplained occurrences often requires expertise from diverse scientific fields. Physicists can contribute by developing theoretical models that account for energy transfers and interactions beyond our current understanding. For instance, exploring the potential for quantum entanglement to play a role in seemingly instantaneous communication across large distances or the implications of higher dimensional physics in explaining apparent violations of causality. Chemists and biologists can analyze physical evidence for unusual chemical reactions or biological anomalies. Psychologists and neuroscientists can investigate the cognitive and neurological factors that may influence subjective perceptions and experiences. This interdisciplinary collaboration ensures that investigations remain comprehensive and avoid the limitations of a single perspective.

Statistical analysis is essential for identifying genuine anomalies from random occurrences. Statistical significance testing is crucial in determining if observed correlations between variables are likely due to chance or represent a real effect. However, the application of statistical methods requires careful consideration of the limitations of existing statistical models and potential biases within data sets. In some cases, a Bayesian approach may be more appropriate, particularly where prior information is available.

Control experiments and blind tests are pivotal to eliminate the influence of experimenter bias and suggestion effects. Double-blind studies, in which neither the subjects nor the experimenters are aware of the experimental conditions, are particularly valuable in minimizing the risk of bias. By systematically controlling variables and introducing placebo effects, we can more accurately assess the validity of observed phenomena. The principle of falsifiability must be upheld; a good scientific hypothesis must be testable and potentially refutable. This ensures that research remains grounded in empirical evidence and avoids the pitfalls of untestable claims.

The ethical implications of investigating these phenomena should be carefully considered. Researchers must prioritize the safety and well-being of all participants and respect the privacy and rights of individuals involved. Any intervention in anomalous situations should be conducted with caution and a thorough risk assessment. This necessitates establishing clear ethical guidelines and ensuring that all research follows strict ethical protocols.

The exploration of anomalous phenomena presents a significant challenge, demanding both intellectual rigor and a willingness to engage with the unknown. A combination of rigorous data collection, advanced instrumentation, interdisciplinary collaboration, and a critical scientific approach is needed. While many questions remain unanswered, a dedicated and scientifically robust methodology offers the best hope for unraveling the mysteries surrounding these events and furthering our understanding of the universe and our place within it. This rigorous pursuit of knowledge, guided by ethical considerations, could lead to breakthroughs that dramatically reshape our understanding of reality itself. The process necessitates patience, perseverance, and a commitment to critical thinking, recognizing that the path to discovery is often fraught with uncertainty and unforeseen challenges. The ultimate goal is not simply to explain these anomalies, but to integrate their understanding into a broader, more comprehensive scientific framework, which may ultimately rewrite the basic laws of physics as we currently understand them. The investigation into these phenomena represents a journey into the unexplored frontiers of human knowledge, pushing the boundaries of our current understanding of the cosmos and our relationship to it.

The previous sections outlined a rigorous, multidisciplinary approach to investigating anomalous phenomena, emphasizing the need for objective data collection, controlled experiments, and interdisciplinary collaboration. However, even with the most meticulous methodologies, we inevitably encounter the limits of scientific explanation. The very nature of "anomalies" implies that they fall outside the established framework of our current scientific understanding. This doesn't invalidate the scientific method; rather, it highlights its inherent limitations and the need for a more nuanced perspective.

One fundamental limitation lies in the inherent biases embedded within the scientific process itself. While rigorous methodologies strive to minimize subjective influences, the very act of choosing which phenomena to investigate, which questions to ask, and which methodologies to employ, inevitably reflects existing paradigms and assumptions. Our scientific models are built upon the data we collect and the questions we deem worthy of investigation. Anomalies, by definition, challenge these pre-existing frameworks, often requiring a paradigm shift to fully accommodate them. The history of science is replete with examples where initial resistance to anomalous findings ultimately led to revolutionary changes in our understanding of the universe – from the acceptance of heliocentricity to the advent of quantum mechanics.

The challenge is further compounded by the inherent limitations of our current scientific instruments and technologies. Many anomalous phenomena manifest at scales or in ways that elude our current detection capabilities. For instance, subtle variations in gravitational fields, hypothesized to be involved in certain psychokinetic phenomena, may lie beyond the sensitivity of current instrumentation. Similarly, the potential for interactions with hypothetical higher dimensions, often invoked in discussions of UFO sightings or altered states of consciousness, remains largely speculative due to the lack of appropriate tools for their detection. The development of new instruments and technologies specifically designed to address these limitations is crucial, though the very nature of these phenomena often makes predicting their characteristics and designing suitable instruments a significant challenge.

Furthermore, many anomalous phenomena are intrinsically subjective, relying on witness accounts and personal experiences. While rigorous documentation and psychological assessment can help mitigate the risk of misinterpretation or fabrication, the inherent subjectivity of these experiences makes them difficult to quantify and analyze using purely objective scientific methods. Near-death experiences, out-of-body experiences, and even certain forms of poltergeist activity often lack readily available physical evidence, making their investigation heavily reliant on witness testimony and subjective accounts. The credibility of such accounts depends heavily on careful evaluation, including consideration of psychological factors, memory biases, and potential social influences. However, dismissing such experiences solely based on the lack of objective evidence ignores the potential richness of human experience and the limitations of current scientific instruments to capture the full spectrum of reality.

Another crucial limitation relates to the vastness and complexity of the universe itself. Our current scientific models, while remarkably successful in explaining many aspects of the cosmos, inevitably represent simplified representations of an infinitely intricate reality. We are only beginning to understand the fundamental forces and constituents of the universe, and the existence of dark matter and dark energy highlights significant gaps in our understanding. The sheer scale of the universe, encompassing unimaginable distances and potentially countless other realities, means that our ability to directly observe and test every phenomenon is inherently limited. Thus, accepting the possibility of phenomena beyond our current understanding isn't a rejection of science, but rather a recognition of its incompleteness and the need for ongoing exploration.

The question of causality also poses significant challenges. Many anomalous phenomena seem to defy conventional notions of cause and effect, suggesting connections between events that are spatially or temporally separated, or exhibiting effects that seemingly precede their causes. These phenomena, often associated with phenomena like precognition or remote viewing, challenge our deeply ingrained assumptions about linearity and causality, necessitating a reevaluation of our fundamental understanding of time and space. The potential role of quantum entanglement, with its seemingly instantaneous correlations between distant particles, offers a tantalizing glimpse into possibilities that extend beyond classical physics, although much work remains to be done to establish a clear connection between quantum phenomena and these purported anomalies.

In the face of these limitations, a degree of open-mindedness is essential. This doesn't imply accepting every claim uncritically, but rather acknowledging the possibility of phenomena that lie beyond our current understanding. It requires a willingness to revise our scientific models in light of new evidence and to explore alternative explanations, even if they challenge established paradigms. The history of science is a testament to the value of open-mindedness, demonstrating how seemingly improbable ideas have been vindicated by subsequent research. The pursuit of understanding anomalous phenomena demands not a rejection of the scientific method, but a refinement of it, acknowledging its limitations and embracing its capacity for ongoing evolution.

The pursuit of knowledge is a journey, not a destination. Our current scientific understanding, for all its achievements, remains incomplete. The study of anomalous phenomena, far from being a distraction from the "real" science, represents an opportunity to expand our understanding of the universe and our place within it. The pursuit of such knowledge requires a combination of rigorous methodology, a willingness to challenge existing paradigms, and a profound respect for the unknown. The limits of scientific explanation, while real, should not be perceived as boundaries, but rather as frontiers waiting to be explored. The mysteries surrounding these anomalies may hold the key to unlocking a deeper, richer understanding of reality, potentially revolutionizing our understanding of physics, cosmology, and even consciousness itself. Embracing this challenge with open minds and rigorous methodologies is not a departure from science, but rather its most profound and exciting extension. It is a testament to the enduring human curiosity and our unwavering quest to unravel the secrets of the universe.

10

The Implications of a Multifaceted Universe

The preceding chapters have laid out a framework for understanding a universe far more multifaceted than our current scientific models allow. We've explored the theoretical underpinnings of density variations creating distinct realms, the potential for parallel universes existing alongside our own, and the intriguing possibility of interdimensional interactions explaining some seemingly paranormal phenomena. But what are the implications of this drastically altered cosmological perspective? How does this re-evaluation of reality shift our understanding of our place within the cosmic tapestry?

The most immediate consequence is a profound humbling. For millennia, humanity has viewed itself as the central actor in a cosmic drama, a species uniquely blessed with consciousness and intelligence. The geocentric model of the universe, though long superseded, reflects a deep-seated anthropocentric bias. Even the heliocentric model, while placing Earth within a larger solar system, still granted humanity a central role within our local cosmic neighborhood. The implications of a multiverse, however, drastically diminish this self-importance. If countless other universes exist, each potentially with its own unique physical laws and life forms, our own universe and our own existence become but a single, infinitesimally small point in an incomprehensibly vast cosmic ocean. We are no longer the center of attention, but merely one tiny ripple in an infinite expanse of possibility.

This humbling, however, is not necessarily a negative one. It allows us to approach the universe with a renewed sense of wonder and curiosity. Instead of clinging to anthropocentric notions, we can embrace the sheer scale and complexity of the cosmos, acknowledging the existence of possibilities far beyond our current comprehension. This new perspective fosters a deeper appreciation for the delicate balance that allows life to flourish in our own universe, a balance that might be exceptionally rare, even unique, in the grand scheme of things. Our existence, though statistically insignificant in the multiverse, becomes all the more precious and remarkable precisely because of its precariousness.

The implications extend beyond a philosophical reassessment of our place in the cosmos. The possibility of other universes, with different physical laws and conditions, opens up exciting new avenues for scientific exploration. The study of cosmology is no longer limited to our observable universe; instead, it encompasses the entire multiverse, presenting a vastly expanded landscape of research opportunities. It compels us to rethink fundamental physical constants, such as the speed of light or the gravitational constant, understanding them not as immutable laws of nature, but as parameters that might vary across different universes. This opens the door to exploring new physics, potentially resolving some of the lingering mysteries that plague our current models, such as the nature of dark energy and dark matter.

Furthermore, the possibility of other universes, even those with vastly different physical laws, challenges our understanding of consciousness itself. If consciousness can arise in one universe, why not in others? The question of whether consciousness is a unique property of our universe, or a more fundamental aspect of reality that transcends specific physical laws, becomes a crucial area of inquiry. The study of consciousness becomes intertwined with the study of the multiverse, offering a fascinating interdisciplinary perspective on one of the most profound mysteries of existence. This interdisciplinary approach might even yield unexpected answers to the hard problem of consciousness, moving us from speculation to empirical investigation. Could the anomalies we've discussed, phenomena seemingly defying our understanding of causality and physics, be hints of interactions between our universe and others?

Moreover, the possibility of interdimensional travel or communication, though currently relegated to the realm of science fiction, becomes a legitimate, albeit highly speculative, area of scientific inquiry. If different realms or dimensions exist, coexisting with our own, the possibility of interacting with them, either through technological means or through altered states of consciousness, must be considered. This raises profound ethical questions about potential interactions with extraterrestrial or interdimensional beings, underscoring the need for careful consideration of the potential consequences of such encounters. The scientific community must establish protocols and guidelines to prevent unintended consequences should such interactions ever become a reality.

Our understanding of time and space is also fundamentally challenged. The multiverse concept suggests a cosmic landscape far more complex than our current models of space-time. The limitations of our current understanding of time may be due to our confinement within a specific universe or dimension, with the true nature of time being far more fluid and dynamic across the multiverse. The seemingly rigid linearity of cause and effect might break down in the context of interdimensional interactions, potentially explaining some of the anomalous phenomena discussed in previous chapters. The concept of parallel universes suggests that the past, present, and future might not be as distinct as we perceive them, raising profound implications for our understanding of causality and free will.

Beyond these scientific implications, the existence of a multiverse profoundly impacts our philosophical and theological perspectives. If our universe is just one of many, the uniqueness of humanity, and even the existence of a creator god, are called into question. The implications are vast, potentially reshaping our understanding of purpose, meaning, and our place within the greater cosmic order. The impact on established religions and belief systems would be seismic, requiring a re-evaluation of fundamental doctrines and beliefs. The exploration of this complex interplay between science and religion is crucial to ensure a smooth and reasoned response to this potentially paradigm-shifting discovery.

However, these far-reaching implications should not overshadow the importance of rigorous scientific methodology. While the concept of a multiverse remains speculative, we must continue to seek evidence, refining our observational tools, and developing new theoretical frameworks to better understand this vast cosmic landscape. The pursuit of knowledge must remain central to our investigation, and the speculative nature of the multiverse must never detract from the importance of methodical, empirical investigation. This requires continued investment in both theoretical and experimental physics, cosmology, and the exploration of those phenomena that currently evade our full comprehension.

In conclusion, the implications of a multifaceted universe are profound and multifaceted, extending far beyond the realm of theoretical physics. It necessitates a re-evaluation of our place in the cosmos, a reassessment of our understanding of physical laws, and a rethinking of our philosophical and theological perspectives. While the multiverse remains a speculative concept, the very possibility demands that we broaden our scientific horizons, embracing the inherent uncertainties of exploring the unknown. This journey into the unknown is not a departure from science but rather its most exciting extension. The exploration of the multiverse, and the potential for other realities and dimensions, promises to be one of the most compelling and transformative scientific endeavors of the coming decades, potentially reshaping our understanding of reality itself.

The implications of a multiverse extend far beyond the purely scientific realm, impacting deeply upon our philosophical understanding of existence. The most immediate and profound shift is a radical decentering of humanity. For centuries, anthropocentric views have dominated our understanding of the cosmos. We've seen ourselves as the pinnacle of creation, the sole possessors of consciousness within a universe designed, perhaps, specifically for our benefit. The Copernican revolution, shifting Earth from the center of the solar system, was a significant blow to this anthropocentric worldview, but the multiverse hypothesis delivers a far more devastating, yet ultimately liberating, blow.

In a multiverse scenario, our universe, with its specific physical constants and laws, is merely one among countless others. Our planet, our sun, our galaxy – all become statistically insignificant blips within a cosmic ocean of unimaginable scale. The sheer probability of other universes harboring life, perhaps even intelligent life vastly exceeding our own capabilities, becomes overwhelmingly high. This realization, while potentially unsettling, can also be deeply empowering. It compels us to relinquish the hubris of self-importance and embrace a profound humility in the face of cosmic vastness. We are not the center, nor are we necessarily unique. This does not diminish our value, but rather contextualizes it within a grander narrative.

Consider the implications for the concept of "uniqueness." If countless universes exist, each with potentially unique physical laws and life forms, the very notion of "uniqueness" becomes relative. What constitutes a unique event or phenomenon within our universe might be commonplace elsewhere. A unique evolutionary pathway on Earth, for instance, might be replicated countless times across other universes, each with subtly different variations. This perspective forces us to reassess our understanding of rarity and probability. What appears extraordinary and improbable from our limited perspective might be remarkably common in the broader context of a multiverse.

This re-evaluation extends to the very nature of consciousness. If consciousness, as we understand it, exists in our universe, why wouldn't it emerge in others? The multiverse hypothesis suggests that consciousness might be a fundamental aspect of reality, not a unique property confined to our particular physical laws. This opens up thrilling avenues for philosophical inquiry. Is consciousness a cosmic constant, a fundamental property of the universe, like energy or matter? Or is it a product of specific conditions, making its emergence statistically rare, even miraculous? Understanding this could unlock profound insights into the nature of reality, linking cosmology, physics, and philosophy in an unprecedented way.

The implications for the philosophy of mind are equally profound. The hard problem of consciousness, the challenge of explaining how physical processes give rise to subjective experience, may find new avenues for resolution within a multiverse framework. Perhaps the subjective experience we call consciousness is not only a function of specific brain states, but also interacts with other universes, or other dimensions within our own. Exploring this intersection of consciousness, physics, and cosmology could revolutionize our understanding of the mind-body problem. Might anomalous experiences, often attributed to the paranormal, be a manifestation of such interdimensional interactions? This is a radical departure from conventional materialism, but a worthy avenue of scientific and philosophical exploration.

Furthermore, the concept of a multiverse forces us to reconsider our understanding of time and space. Our current models of spacetime might be severely limited by our confinement within a single universe. The true nature of time and space might be far more fluid and dynamic across the multiverse, potentially defying our linear perceptions of cause and effect. Could parallel universes, existing simultaneously, influence each other, leading to phenomena that defy our conventional understanding of causality? This suggests that our perception of reality may be a product of our limited perspective, and the true nature of spacetime might be far more interconnected and complex than we currently fathom. Our understanding of free will, for example, could be dramatically altered if the past, present, and future are not as rigidly defined as we assume.

The philosophical impact on theology and religion is equally significant. The existence of a multiverse potentially undermines many traditional theological frameworks that posit a unique creation or a specific divine plan for humanity. If our universe is merely one among countless others, the questions of creation, purpose, and the existence of a creator God must be re-examined. However, this does not necessarily lead to atheism or a rejection of spirituality. Instead, it prompts a reassessment of traditional religious doctrines and a potential shift towards more inclusive and expansive theological perspectives. The multiverse hypothesis might even inspire new forms of spirituality, emphasizing the interconnectedness of all existence and the vastness of cosmic possibility.

The ethical implications are also profound. If interdimensional travel or communication becomes feasible, a whole new realm of ethical considerations arises. How should we interact with potential inhabitants of other universes or dimensions? What are our responsibilities to other forms of life, perhaps drastically different from anything we can currently imagine? Developing ethical frameworks to govern such interactions is a critical task, requiring careful consideration of potential risks and benefits. The scientific community must collaborate with ethicists, philosophers, and theologians to establish guidelines that promote responsible exploration and prevent unintended consequences. Our understanding of humanity's place within the cosmos is not simply altered; it is utterly redefined, forcing us to confront fundamental questions about our values and responsibilities.

In conclusion, the philosophical implications of a multifaceted universe are vast and complex, pushing the boundaries of our current understanding of existence, consciousness, and the nature of reality. This is not merely a scientific endeavor but a profound philosophical and spiritual journey, demanding a re-evaluation of our fundamental beliefs and values. The multiverse hypothesis, even in its speculative stage, compels us to embrace uncertainty, to foster interdisciplinary collaboration, and to confront the inherent limitations of our anthropocentric perspectives. The journey of exploring the possibilities, the potential, and the profound implications of this cosmic landscape is far from over, and it promises to be one of the most transformative intellectual endeavors in human history.

The implications of a multiverse profoundly reshape our theological landscape, challenging long-held beliefs and opening up new avenues for spiritual exploration. For centuries, religious narratives have often presented a singular, meticulously crafted cosmos, with humanity occupying a central, divinely ordained role. The concept of a multiverse, however, presents a universe – or rather, universes – of vastly different scales and possibilities, prompting a critical re-evaluation of creation myths, divine plans, and the very nature of divinity itself.

One of the most immediate impacts is on creation narratives. Many religions describe a single act of creation, a singular beginning from which all things emanate. The multiverse hypothesis, however, suggests a far more complex, potentially infinite, series of origins. If our universe is just one among countless others, each with its own unique physical laws and potentially its own genesis events, the traditional narratives of creation require significant reinterpretation. This doesn't necessarily invalidate the core tenets of faith for believers; instead, it may necessitate a more nuanced and expansive understanding of divine action. The creator, rather than being responsible for a single, meticulously planned universe, could be viewed as the architect of a vast, multifaceted cosmic tapestry, each thread a unique universe with its own story.

This expanded perspective on creation also has implications for the concept of divine providence. If a deity oversees an infinite multiverse, the level of direct intervention in each individual universe might become statistically insignificant. The concept of a meticulously planned trajectory for humanity, for instance, may seem less plausible in a context where countless other universes may harbor different forms of intelligent life, each with their own evolutionary paths and destinies. This does not necessarily diminish the significance of divine guidance or purpose; instead, it may necessitate a shift towards a more indirect or overarching form of divine interaction. Perhaps divine purpose is not confined to a single universe but extends across the entire multiverse, encompassing a vast and incomprehensible cosmic plan.

The question of the nature of divine consciousness itself takes on new dimensions within a multiverse framework. If our universe is just one among many, is there a single, unified divine consciousness encompassing all realities, or are there multiple, perhaps even interacting, divine entities, each responsible for a subset of universes? This opens up fascinating avenues for theological speculation, challenging traditional monotheistic and polytheistic models alike. It may also lead to new forms of pantheism, where the divine is not separate from but inherent within the very fabric of the multiverse. The ultimate reality could be an incomprehensible unity encompassing all existence, defying our limited human comprehension.

Consider the implications for the problem of evil. Traditional theodicies, which attempt to reconcile the existence of a benevolent God with the presence of suffering in the world, may require significant revision within a multiverse context. If our universe is merely one among many, and each universe has its own unique physical laws and evolutionary trajectories, the reasons for suffering in our universe might be far more complex and less directly attributable to divine will than previously assumed. The existence of other universes, some perhaps far more idyllic than our own, could suggest that suffering is not a necessary feature of all creation, but a contingent aspect of specific universes with specific properties. This does not necessarily resolve the problem of evil, but it may provide a more nuanced and sophisticated framework for considering it.

The relationship between science and religion also undergoes a profound transformation in a multiverse context. For centuries, a perceived conflict between scientific findings and religious beliefs has fueled much debate. The multiverse hypothesis, however, opens up the possibility of a more harmonious coexistence. Science may explore the physical properties of the multiverse, while religion might focus on the spiritual, philosophical, and ethical dimensions of existence within this vast cosmic landscape. The multiverse is not simply a scientific concept; it's a profound metaphysical reality, touching upon questions of meaning, purpose, and the nature of reality itself.

Furthermore, the impact on various religious traditions would be unique. In some interpretations, the multiverse could reinforce existing beliefs. For instance, the concept of multiple heavens or realms in some religions could find resonance with the idea of multiple universes. Certain mystical traditions that emphasize the interconnectedness of all things may find a natural alignment with the concept of a multiverse, viewing each universe as a facet of a greater cosmic unity. However, other traditions might face a greater challenge in integrating the multiverse hypothesis into their existing frameworks. A careful and nuanced approach is necessary to accommodate the perspectives of various religious traditions, recognizing that the interpretation of the multiverse within a theological context will vary widely.

Finally, the multiverse hypothesis encourages a broader, more inclusive approach to spirituality. By recognizing the vastness and complexity of cosmic reality, we might become less focused on narrow, parochial perspectives and more receptive to the diversity of spiritual experiences and perspectives. The multiverse encourages humility and respect for the unknown, fostering an attitude of open-mindedness and intellectual curiosity. This new perspective can lead to a deeper appreciation for the mysteries of existence and a more profound sense of wonder at the vastness of the cosmos and the potential for existence beyond our limited understanding. The theological implications of a multiverse are profound and multifaceted, prompting a reevaluation of our beliefs, expanding our understanding of the divine, and ultimately enriching our spiritual journeys. It is a journey of exploration that requires open-mindedness, interdisciplinary collaboration, and a willingness to engage with the profound mysteries that lie at the heart of existence.

The ethical implications of a multifaceted universe are profound and far-reaching, extending beyond mere scientific curiosity into the realm of moral responsibility. If other dimensions or realities exist, as the preceding chapters have explored, then our actions within our own reality could have unforeseen and potentially devastating consequences for these other planes of existence. The very act of interaction, even if unintentional, could disrupt delicate balances, alter the course of evolution in other universes, or even lead to the annihilation of entire civilizations. This necessitates a cautious and deeply considered approach to any potential contact or interaction with other dimensions, guided by a robust ethical framework that prioritizes the well-being of all sentient life, wherever it may reside.

One of the primary ethical concerns revolves around the potential for exploitation. If we were to gain access to other dimensions, there's a significant risk of exploiting their resources or inhabitants for our own benefit. Imagine the potential for harvesting energy from another universe, a universe where the laws of physics might allow for far more efficient energy generation than our own. Such an action, even if seemingly beneficial for humanity, could be catastrophic for the beings inhabiting that universe, potentially disrupting their ecological balance or even causing their extinction. This echoes historical instances of human exploitation of Earth's resources and other life forms, demonstrating a persistent pattern of self-serving behavior that must be consciously challenged and overcome when considering interactions with other realities.

Furthermore, the potential for introducing diseases or harmful substances across dimensional boundaries presents a serious ethical hazard. Our own biological and chemical makeup might be incompatible with the life forms of other universes, leading to devastating pandemics or environmental catastrophes. The simple act of travelling between dimensions could introduce unknown pathogens or pollutants that could irrevocably alter the delicate balance of these alien ecosystems. This highlights the need for rigorous quarantine protocols and comprehensive research before any contact is attempted, mirroring the precautionary measures undertaken to prevent the introduction of terrestrial diseases to other planets in science fiction scenarios. The scale of potential harm is dramatically amplified when considering entire universes.

The question of consent also arises prominently in the ethical considerations of interacting with other dimensions. How can we ethically interact with beings who might not possess the capacity to comprehend or consent to our actions? If we encounter intelligent life in another dimension, what is our responsibility to ensure their autonomy and self-determination? Our interactions must be guided by the principle of informed consent, ensuring that any interaction is mutually beneficial and respectful of the other party's rights and choices. This might necessitate developing sophisticated interdimensional communication technologies and employing ethical protocols to ensure that any interaction is voluntary and truly informed. We must be aware of the immense power imbalance inherent in such interactions, and strive to level the playing field as much as possible.

The concept of "non-interference" becomes paramount in this context. Just as many ethical codes advocate for minimizing human intervention in natural ecosystems, a similar principle must guide our interaction with other universes. We must acknowledge the possibility that these other realities have evolved complex ecological systems and societies with their own intrinsic value, and that our intrusion could have unpredictable and devastating repercussions. Interfering in their natural processes, even with the intention of helping, could have unintended consequences, leading to the disruption of delicate balances and possibly the destruction of entire civilizations. Respecting the autonomy and integrity of these other universes becomes a cornerstone of responsible interdimensional interaction.

The ethical considerations extend beyond direct interaction to encompass the indirect consequences of our actions. Even seemingly innocuous research into other dimensions could have significant implications. The accumulation of knowledge about other universes could be misused for malicious purposes, leading to the development of weapons or technologies capable of destroying entire realities. This highlights the need for careful regulation and ethical oversight of interdimensional research, ensuring that knowledge is used responsibly and not exploited for destructive ends. This responsibility extends to the public as well, necessitating a responsible and ethically guided discourse regarding the nature and potential of other universes.

Moreover, the potential impact on our own society should also be considered. The discovery of other dimensions or realities might lead to societal upheaval, religious crises, or even widespread panic. The psychological impact of such a discovery could be profound, challenging fundamental assumptions about humanity's place in the cosmos and potentially destabilizing existing social structures. Preparation for such an eventuality requires a multifaceted approach, including public education, psychological support systems, and the development of mechanisms to manage societal response. A responsible and ethical approach mandates a concerted effort to mitigate any negative societal consequences that may arise from the exploration and interaction with other dimensions.

Furthermore, the very act of acknowledging the possibility of other dimensions and realities shifts our perspective on the concept of "humanity." If we are not alone in the cosmos, but rather a part of a much larger multiverse, our understanding of our identity and place in the universe undergoes a fundamental transformation. This requires a re-evaluation of our ethical frameworks, expanding them beyond anthropocentric concerns to encompass a wider cosmological perspective. A true ethical compass for interdimensional exploration must consider the needs and rights of all sentient beings, regardless of their origin or the universe they inhabit.

Finally, the ethical implications of a multifaceted universe highlight the profound responsibility that comes with scientific advancement. As our understanding of the cosmos expands, so too does our capacity to impact other realities, both positively and negatively. The exploration of other dimensions should not be driven solely by scientific curiosity, but guided by a deep sense of ethical responsibility. This necessitates a collaborative effort between scientists, ethicists, theologians, and policymakers to develop comprehensive guidelines and protocols for interdimensional interaction. Only through careful consideration and a commitment to responsible behavior can we ensure that our interactions with other realities contribute to the greater good of the cosmos as a whole. The future of our interaction with other dimensions rests upon our ability to prioritize ethical considerations above scientific curiosity, recognizing that the consequences of our actions could extend far beyond the confines of our own known reality. This necessitates a global, interdisciplinary approach, combining scientific rigor with ethical sensitivity and a deep respect for the potential wonders and dangers that may lie beyond our current understanding.

The ethical considerations outlined in the preceding sections lay the groundwork for a crucial next step: charting the course of future research into the multifaceted universe. This endeavor demands a paradigm shift, moving beyond the limitations of conventional scientific methodologies and embracing a more holistic, interdisciplinary approach. The sheer complexity of the challenge necessitates a collaborative effort involving physicists, cosmologists, parapsychologists, ethicists, theologians, and philosophers, among others. This interdisciplinary dialogue is not merely desirable; it is essential for navigating the uncharted territories of other realities responsibly.

One of the most pressing research needs is the development of new theoretical frameworks and mathematical models capable of describing and predicting the behavior of these other realities. Current physics, while remarkably successful in explaining our own universe, may prove inadequate for the task. We require novel theoretical constructs that can account for variations in fundamental constants, different dimensions, and potentially, alternative laws of physics. String theory, loop quantum gravity, and other theoretical advancements offer potential starting points, but significant further development and refinement are necessary. This requires investment in fundamental theoretical research, providing the intellectual scaffolding upon which future experimental investigations can be built.

The development of new experimental methodologies is equally critical. Direct observation of other dimensions or realities remains a significant challenge. While indirect evidence, such as anomalous phenomena, might suggest their existence, concrete proof requires sophisticated and innovative approaches. One promising avenue is the further development of advanced sensing technologies, particularly in areas such as quantum entanglement and gravitational wave detection. These technologies could potentially allow us to detect subtle variations in spacetime or energy signatures indicative of interactions with other realities. Furthermore, research into advanced computing and artificial intelligence could play a significant role in analyzing large datasets and identifying patterns that might otherwise go unnoticed. This requires a substantial investment in the development of dedicated experimental facilities and computing infrastructure, specifically designed for interdimensional research.

Beyond the technological advancements, a critical area for future research lies in the development of reliable methods for interdimensional communication. If other realities harbor sentient life, establishing contact would revolutionize our understanding of the universe. However, this necessitates overcoming the considerable challenges inherent in communicating across potentially vast physical distances and differing physical laws. Research into quantum communication, advanced telepathy research (within the framework of rigorous scientific methodology), and the study of non-local phenomena could offer pathways toward achieving interdimensional communication. This demands not only technological advancements but also a deeper understanding of consciousness, its nature, and its potential role in bridging the gap between realities.

The study of anomalous phenomena—events defying conventional scientific explanation, such as UFO sightings, psychic abilities, and unexplained energy anomalies—deserves renewed attention within a rigorous scientific framework. While these phenomena are often dismissed as fringe science, their persistent occurrence warrants serious investigation, particularly given the possibility that they might represent interactions with other realities. This necessitates developing robust methodologies for investigating these phenomena, minimizing biases, and ensuring data integrity. A systematic and scientifically rigorous approach to studying anomalous phenomena could provide valuable insights into the nature of the multifaceted universe, potentially revealing clues about the underlying mechanisms governing the interaction of different realities.

Furthermore, future research must address the ethical implications of interacting with other realities. The development of ethical guidelines and protocols is not merely an academic exercise but a crucial step in preventing unintended consequences. This necessitates interdisciplinary collaboration between scientists and ethicists, leading to the establishment of international organizations or agreements that govern interdimensional research and contact. The establishment of such ethical frameworks should prioritize the safety and well-being of all sentient beings, whether in our own reality or others. This collaborative effort will be pivotal in ensuring that our interactions with other realities are guided by a strong moral compass.

In addition to the scientific and ethical challenges, future research must also consider the sociological and psychological impacts of discovering other realities. The societal ramifications of such a discovery could be profound, ranging from widespread societal upheaval to religious crises. Preparation for this eventuality requires a multifaceted approach, involving public education, the development of appropriate psychological support systems, and strategies for managing potential societal unrest. This requires research in social psychology and sociology to better understand and anticipate the potential societal responses to the discovery of other realities, thereby creating a framework for managing these potential societal disruptions.

Moreover, future research needs to address the potential for unintended consequences of our actions on other realities. This includes the development of robust safety protocols and the creation of an interdimensional quarantine system to prevent the accidental introduction of harmful substances, diseases, or technologies. This is not merely a precaution; it is a necessity to prevent catastrophic outcomes for both our own reality and any other realities we might encounter. This research should incorporate principles of risk assessment and disaster management to assess potential risks and develop appropriate mitigation strategies.

The integration of different disciplines is fundamental to future research directions. The exploration of other realities requires a concerted effort involving theoretical physicists, experimental physicists, cosmologists, parapsychologists, ethicists, theologians, social scientists, and experts in other fields. The creation of interdisciplinary research centers and collaborative networks would facilitate the exchange of ideas and the development of innovative research strategies. Such collaboration will enable researchers to share expertise, leverage their respective methods, and develop holistic, comprehensive approaches to the study of the multifaceted universe.

Finally, future research must embrace a global perspective. The discovery of other realities is not a matter of national interest alone; it is a concern for all of humanity. International collaboration and the establishment of global ethical standards are essential to prevent misuse of the knowledge and technologies developed in this field. This calls for international scientific cooperation, the establishment of global ethical frameworks, and the promotion of an open and transparent dialogue to discuss the implications of this research. This necessitates the establishment of international committees and bodies to ensure responsible development and application of any relevant technologies and knowledge. Only through a concerted global effort can we ensure that our exploration of the multifaceted universe benefits all life, present and future, across all realities. The path forward requires not only scientific advancement but a commitment to responsible stewardship of our universe and all the others that may share the cosmos with us.

11

The Science of the Paranormal

The exploration of paranormal phenomena necessitates a rigorous and systematic approach, moving beyond anecdotal evidence and subjective interpretations. To engage in meaningful scientific inquiry, we must first establish clear definitions and categories for the phenomena under investigation. This is not to say that all reported events are necessarily genuine or explainable within a current scientific paradigm, but rather to provide a structured framework for analysis, allowing for the potential identification of patterns and underlying mechanisms. The sheer diversity of reported paranormal events necessitates a hierarchical classification system.

One primary categorization hinges on the nature of the purported interaction: *physical* versus *mental*. Physical paranormal phenomena involve interactions with the physical world that defy conventional scientific understanding. This category encompasses a wide range of events, including poltergeist activity (unexplained movements of objects), spontaneous human combustion (the purported self-ignition of a human body), apparitions (visual manifestations of seemingly disembodied entities), and various forms of unexplained energy anomalies. The challenge lies in differentiating these phenomena from mundane explanations, such as misperceptions, hoaxes, or unknown natural processes. Rigorous investigation often requires detailed analysis of physical evidence, witness testimony, and environmental factors. Technological advancements, such as advanced sensor technology and sophisticated data analysis techniques, play a critical role in providing objective evidence to corroborate or refute reported events.

The study of physical phenomena often intersects with the field of parapsychology, which seeks to investigate alleged paranormal abilities, often grouped under the umbrella term of *psi* phenomena. This encompasses *psychokinesis* (PK), the purported ability to influence physical systems with the mind, and *telekinesis*, a related term often used interchangeably with PK but sometimes implying a more forceful or direct influence. While rigorous experimental protocols are crucial to minimize bias and ensure data integrity, these phenomena pose significant challenges to current physical theories, requiring the development of new models to explain how consciousness might interact with matter. Exploring this interface requires interdisciplinary collaborations, bringing together expertise in physics, neuroscience, and psychology.

Another broad category focuses on *mental* paranormal phenomena, encompassing alleged abilities to access or influence information beyond the constraints of normal sensory perception or cognitive processes. *Clairvoyance*, the purported ability to perceive information beyond normal sensory channels, and *precognition*, the purported ability to perceive future events, are key examples. *Telepathy*, the purported ability to transmit or receive thoughts or emotions without the use of conventional communication channels, falls under this category as well. These phenomena present unique challenges for investigation, as their subjective nature necessitates robust experimental designs that control for bias and deception. The use of double-blind studies, independent verification of results, and rigorous statistical analysis are crucial for establishing credibility. However, the inherent difficulties in quantifying and objectively measuring mental processes remain a significant hurdle.

Within these broader categories, numerous subcategories exist, reflecting the specific nature of the reported events. For instance, *haunted locations* often exhibit a combination of physical and mental phenomena, including unexplained noises, apparitions, and sensations of being watched. The interpretation of such events often relies heavily on cultural context and belief systems, adding another layer of complexity to their investigation.

The analysis of anecdotal reports must be approached with caution. While personal testimonies can provide valuable starting points for investigation, they are inherently susceptible to biases such as confirmation bias, suggestibility, and memory distortion. Furthermore, the inherent subjectivity of many paranormal experiences makes objective verification extremely difficult. This highlights the crucial need for rigorous scientific methodologies, emphasizing controlled experiments, detailed data collection, and statistical analysis to ensure objectivity and minimize bias. In many cases, a thorough investigation often reveals mundane explanations, highlighting the importance of eliminating plausible alternative hypotheses before attributing phenomena to paranormal causes.

Moreover, the challenge of replicability is paramount. A hallmark of genuine scientific phenomena is their ability to be replicated under controlled conditions. The difficulty in replicating many purported paranormal events casts doubt on their validity as genuine scientific phenomena. However, this should not necessarily lead to their dismissal, but rather to a careful consideration of the limitations of current experimental methods and the development of novel approaches that are sensitive enough to detect subtle or intermittent effects. It may be the case that the underlying mechanisms responsible for some paranormal phenomena are intermittent, sensitive to specific environmental factors, or dependent on the subtle interaction between consciousness and the physical world.

Furthermore, the distinction between genuine paranormal events and fraudulent activities must be carefully considered. The history of paranormal investigation is replete with documented instances of deliberate deception, ranging from simple stage magic tricks to elaborate hoaxes. This underscores the necessity of rigorous scrutiny of evidence, thorough background checks of witnesses, and the application of forensic techniques to detect potential manipulation or fabrication. In some cases, sophisticated methods of deception may have misled even trained investigators, emphasizing the importance of continuous critical evaluation and the ongoing development of more advanced investigative techniques.

The exploration of paranormal phenomena necessitates a multidisciplinary approach, drawing upon the expertise of physicists, cosmologists, psychologists, neuroscientists, anthropologists, and historians. Each discipline offers unique perspectives and methodologies that can contribute to a more comprehensive understanding of these events. Furthermore, the ethical considerations associated with paranormal research must be explicitly addressed. Issues of informed consent, protection of participants, and responsible dissemination of findings must be carefully managed to ensure the integrity of the research and the well-being of all involved.

The challenge of studying paranormal phenomena lies not just in the inherent difficulty of the subject matter but also in the cultural and societal context within which these phenomena are interpreted. The strong beliefs and emotional investments that surround these phenomena often hinder objective scientific inquiry. This highlights the necessity of fostering open-mindedness and a critical, analytical approach within the scientific community, balancing scientific rigor with a willingness to explore unconventional hypotheses. The potential rewards of successfully unraveling the mysteries surrounding paranormal events are immense, potentially revolutionizing our understanding of the universe and our place within it. This will require a sustained, collaborative effort between scientists, researchers, and the broader community, promoting a culture of open dialogue, critical analysis, and ethical responsibility. The path forward will require embracing both skepticism and intellectual curiosity, pursuing a balance that fosters both the rigorous examination of evidence and the courageous exploration of the unknown.

The application of rigorous scientific methodology to the study of paranormal phenomena presents a formidable challenge, stemming from the very nature of the phenomena themselves. Unlike established scientific fields where repeatable experiments and controlled variables are the norm, paranormal events are often characterized by their unpredictability, rarity, and subjective nature. This inherent variability makes it exceedingly difficult to establish robust, statistically significant results. Furthermore, the lack of a universally accepted theoretical framework for explaining these phenomena hampers the development of standardized experimental designs. Existing scientific instruments and methodologies, honed over centuries for the study of predictable physical processes, often prove inadequate for capturing the subtlety and complexity of purported paranormal interactions.

One major hurdle is the problem of measurement. Many paranormal phenomena, such as telepathy or precognition, involve subjective experiences that are difficult to quantify objectively. While self-reported experiences can provide valuable data, they are inherently susceptible to biases, including recall bias, confirmation bias, and demand characteristics. Participants might subconsciously want to believe in the phenomenon, or they might inadvertently influence their responses to fit what they perceive as the expected outcome. To mitigate these issues, researchers often employ double-blind protocols, where neither the participants nor the experimenters know the conditions of the experiment, and rigorous statistical analysis to control for chance occurrences. However, even with these safeguards, the inherently subjective nature of the data remains a significant challenge.

Another significant issue is the lack of replicability. A cornerstone of scientific validity is the ability to replicate findings under controlled conditions. The difficulty in replicating many purported paranormal events has frequently been cited as evidence against their validity. However, it is crucial to consider that the underlying mechanisms responsible for such phenomena might be inherently unpredictable or contingent upon specific environmental factors or subtle psychological states that are currently poorly understood and thus difficult to reproduce. The inability to replicate an event does not necessarily negate its occurrence; rather, it points towards the limitations of our current methods and the need for more sophisticated experimental designs and more sensitive measurement techniques.

The development of new instrumentation and measurement techniques is crucial. Current technologies, while highly advanced in many respects, may not be sufficiently sensitive to detect subtle paranormal effects. For example, the detection of subtle energy fields associated with alleged poltergeist activity or psychic phenomena might require the development of entirely new sensor technologies. Similarly, measuring subtle changes in brain activity correlated with precognitive experiences requires advanced neuroimaging techniques with high temporal and spatial resolution, pushing the boundaries of current neuroscience methodologies. Furthermore, the exploration of novel data analysis techniques is crucial, utilizing sophisticated algorithms and machine learning approaches to identify patterns and correlations within complex datasets that might elude conventional statistical analysis.

Beyond the challenges of instrumentation and measurement, the interpretation of results presents further complexities. Even with rigorous experimental design and advanced technology, the interpretation of results often requires a delicate balance between skepticism and open-mindedness. Researchers must carefully consider alternative explanations, including mundane factors such as coincidence, suggestion, or experimenter bias. The application of Occam's Razor—the principle that the simplest explanation is usually the best—is crucial in evaluating potential paranormal claims. However, it is equally important to avoid prematurely dismissing unusual findings based solely on a lack of readily available explanation. The discovery of genuinely novel phenomena often requires a willingness to consider hypotheses that initially seem unconventional or even implausible.

Furthermore, the ethical considerations surrounding paranormal research are paramount. Participants must be fully informed about the nature of the study and give their informed consent. Researchers must take steps to protect participants from potential harm, both physical and psychological. The responsible dissemination of findings is also crucial, avoiding sensationalism and ensuring that results are presented accurately and transparently. The ethical challenges are particularly acute when dealing with vulnerable populations, such as individuals who are experiencing grief or trauma, and who might be more susceptible to suggestions or exploitation.

The exploration of paranormal phenomena necessitates a multidisciplinary approach. A collaborative effort involving physicists, cosmologists, psychologists, neuroscientists, and anthropologists is essential to address the complex issues involved. Physicists can bring their expertise in measurement and instrumentation, while psychologists can contribute their knowledge of human cognition and perception. Neuroscientists can provide insights into brain function and consciousness, while anthropologists can offer valuable perspectives on cultural beliefs and practices related to paranormal experiences. A holistic approach, integrating these diverse perspectives, is likely to yield the most significant progress.

Finally, the study of paranormal phenomena must be grounded in a spirit of intellectual humility. The complexities involved are immense, and our current understanding is limited. Acknowledging this limitation is not a sign of weakness but rather a necessary step towards meaningful progress. It is essential to approach this field with both rigorous skepticism and a willingness to consider the possibility of phenomena that lie beyond the current boundaries of established science. The journey towards understanding these enigmatic occurrences is a long and challenging one, but the potential rewards—a deeper understanding of the universe and our place within it—are immeasurable. The path forward requires a sustained commitment to rigorous scientific methodology, technological innovation, ethical research practices, and a willingness to explore the unknown with open minds and critical intellect. Only through such a persistent and comprehensive approach can we hope to unravel the mysteries surrounding paranormal phenomena and possibly revolutionize our understanding of reality itself.

The application of statistical analysis to paranormal claims is crucial for separating genuine anomalies from the noise of coincidence and bias. While the inherent difficulties in studying these phenomena remain significant, statistical methods offer a powerful tool for quantifying the strength of evidence and assessing the probability that observed results are due to chance. The most basic approach involves calculating the probability of a specific event occurring by random chance. For instance, consider a series of successful attempts at remote viewing, where participants accurately describe a target location without any prior knowledge. By calculating the probability of achieving this level of accuracy purely through guessing, we can establish a baseline for evaluating the significance of the results. If the probability of achieving the observed results by chance is exceedingly low (e.g., less than 0.05, a commonly used threshold in scientific research), it suggests that alternative explanations, such as genuine remote viewing abilities, warrant further investigation.

However, the simplicity of this approach is often deceptive. The complexity inherent in many paranormal experiments necessitates more sophisticated statistical methods. For example, in studies of psychokinesis (the ability to influence physical systems with the mind), researchers might use techniques such as analysis of variance (ANOVA) or regression analysis to account for multiple variables that could influence the outcome. These variables might include the participant's psychological state, environmental factors, and the specific experimental apparatus. By incorporating these variables into the statistical model, researchers can more accurately assess the contribution of psychokinesis, if any, to the observed results.

Furthermore, the issue of multiple comparisons needs careful consideration. In many paranormal studies, multiple tests are performed, increasing the likelihood of obtaining a statistically significant result purely by chance. To correct for this, researchers often use procedures such as the Bonferroni correction, which adjusts the significance level to account for the number of comparisons. Ignoring this correction can lead to a significant inflation of Type I errors, meaning that researchers might falsely conclude that a significant effect exists when it does not.

Beyond basic statistical tests, Bayesian methods offer a powerful alternative for evaluating paranormal claims. Bayesian statistics incorporate prior beliefs about the probability of an event occurring, updating these beliefs based on new evidence obtained from the experiment. This approach allows researchers to quantify their uncertainty about the phenomenon under investigation and to update their beliefs as more data becomes available. For example, a Bayesian approach might start with a prior belief that the probability of successful telepathic communication is low, but this belief would be updated based on the results of a well-designed experiment. If the experiment yields strong evidence supporting telepathy, the posterior belief (the updated belief after considering the experimental evidence) will shift towards a higher probability of telepathic communication.

Another crucial aspect of statistical analysis in paranormal research involves dealing with the subjective nature of the data. Many paranormal phenomena, such as near-death experiences or out-of-body experiences, rely on subjective reports rather than objective measurements. While self-reported data can provide valuable insights, it is essential to acknowledge the limitations and potential biases associated with this type of data. Researchers can use qualitative methods, such as thematic analysis, to identify recurring patterns and themes within subjective reports. However, combining these qualitative findings with quantitative analyses, such as statistical modeling of response patterns, strengthens the overall analysis and facilitates a more robust interpretation of the results.

The problem of selective reporting, or the tendency to only report positive results and suppress negative or null results, also poses a significant threat to the validity of paranormal research. This selective reporting can lead to an overestimation of the true effect size, misleading readers and potentially hindering the advancement of the field. To mitigate this issue, researchers should adhere to strict reporting guidelines, including reporting all data, both positive and negative, and describing the methods used for data analysis in detail. Peer review plays an important role in identifying instances of selective reporting and ensuring the transparency of research findings. Moreover, the adoption of pre-registration of studies can significantly help mitigate selective reporting, where researchers specify their hypotheses and analysis plan before collecting any data.

Furthermore, the influence of experimenter bias, where the researcher's expectations unconsciously influence the results, is a major concern in paranormal research. This bias can manifest in various ways, such as the selection of participants, the design of the experiment, and the interpretation of the data. To minimize the impact of experimenter bias, double-blind experiments, where neither the participant nor the experimenter knows the true experimental condition, are essential. Furthermore, the use of automated data collection systems can further reduce the risk of experimenter bias. The adoption of such rigorous control measures significantly enhances the credibility of the findings.

Meta-analysis, a statistical technique that combines the results of multiple studies on the same topic, offers a powerful approach for evaluating the overall strength of evidence for paranormal phenomena. By pooling the data from several independent studies, meta-analysis can increase the statistical power of the analysis and provide a more robust estimate of the effect size. Meta-analyses of paranormal research have yielded mixed results, with some studies showing statistically significant effects, while others fail to find evidence of any paranormal phenomena. The heterogeneity of results across different studies underscores the complexity of the field and the need for more rigorous and standardized methodologies.

Finally, it's crucial to emphasize the limitations of statistical analysis in the context of paranormal research. Statistical significance does not necessarily imply scientific or practical significance. A statistically significant result might still represent a small or insignificant effect in the real world. Conversely, a statistically non-significant result might still be indicative of a genuine phenomenon if the study lacks sufficient power or suffers from other methodological flaws. The interpretation of statistical results in paranormal research must be conducted cautiously, considering the various sources of error and uncertainty. A holistic approach, incorporating both statistical analysis and careful consideration of methodological limitations, is essential for a balanced and informed assessment of paranormal claims. Ultimately, the pursuit of understanding these phenomena necessitates a persistent and comprehensive approach, blending rigorous scientific methodology with an open mind capable of embracing the possibility of phenomena beyond our current understanding.

Controlled experiments, the gold standard of scientific inquiry, have been applied to parapsychological research with varying degrees of success. The challenge lies in designing experiments that are both rigorous enough to withstand scientific scrutiny and sensitive enough to detect subtle effects that might be associated with paranormal phenomena. One significant hurdle is the inherent difficulty in controlling for extraneous variables. Unlike in physics or chemistry where controlled conditions are relatively straightforward to establish, parapsychological experiments often involve human subjects, whose cognitive and emotional states can significantly impact the results.

One prominent area of investigation is psychokinesis (PK), the purported ability to influence physical systems with the mind. Numerous studies have attempted to demonstrate PK effects through experiments involving random number generators (RNGs). RNGs are electronic devices designed to produce sequences of random numbers. Participants are asked to mentally influence the output of these RNGs, aiming to deviate from randomness in a predictable way. While some studies have reported statistically significant deviations from chance expectation, suggesting possible PK effects, others have failed to replicate these findings. The inconsistencies across studies highlight the difficulties in achieving reliable, repeatable results, and the potential for subtle biases to influence outcomes. Moreover, even statistically significant results must be carefully scrutinized to determine whether they represent a genuine phenomenon or are due to methodological artifacts. For example, the possibility of subtle physical influences on the RNGs, undetected flaws in the experimental design, or even subtle biases in the data analysis procedures must be thoroughly investigated and ruled out.

Another critical area involves experiments on telepathy, the supposed ability to transfer thoughts or information between individuals without using the known senses. Classic telepathy experiments often employ techniques like Ganzfeld experiments. The Ganzfeld procedure involves placing participants in a sensory deprivation environment—usually with translucent red light over their eyes and headphones playing white noise—to reduce sensory input and potentially enhance their sensitivity to subtle psychic signals. One participant, the "sender," views a target image while the other, the "receiver," attempts to describe the image based on any mental impressions received. Statistical analysis is then used to determine whether the receiver's descriptions match the target image more often than would be expected by chance. Again, the problem of replication plagues these studies. While some studies have yielded results that suggest potential telepathic effects, others have failed to replicate these results, casting doubt on the reliability of the findings. Crucial methodological considerations include the possibility of subtle sensory leakage, the influence of experimenter bias, and the potential for subjective interpretation of results. A receiver's description might seem vaguely related to the target image but still be far from a definitive match. Careful, rigorous rating scales and objective measures are therefore crucial.

Remote viewing, the alleged ability to perceive information about a distant location without any physical connection, has also been the subject of controlled experiments. In these experiments, a "viewer" attempts to describe a remote location that has been randomly selected by the experimenters. Later, the viewer's descriptions are compared to information about the location. Similar to telepathy experiments, the interpretation of results in remote viewing studies is often subjective and prone to bias. The ambiguity of descriptions makes it difficult to objectively assess whether the viewer's descriptions match the target location to a degree beyond chance. Moreover, the possibility of unconscious cues, such as information inadvertently revealed during the experiment, must be meticulously controlled. Experiments should be conducted under double-blind conditions where neither the viewer nor the judge knows the identity of the target location.

Precognition, the purported ability to perceive future events, presents an exceptionally challenging area for controlled experimentation. Difficulties in defining and objectively measuring future events create significant hurdles. Many studies have attempted to investigate precognition using techniques like testing participants' ability to predict future outcomes of random events. The use of objective, randomly generated sequences of events, recorded prior to the participant's predictions, is essential to prevent bias. The analysis of such experiments requires sophisticated statistical techniques to account for the potential for random fluctuations in results. Furthermore, concerns remain regarding the potential for selection biases – the tendency for researchers to selectively publish or highlight studies with positive outcomes – and the replication crisis inherent in much parapsychological research.

The use of sophisticated statistical methods is crucial in interpreting the results of controlled experiments in parapsychology. However, the reliance solely on statistical significance is insufficient. A statistically significant result, while suggestive, does not conclusively prove the existence of a paranormal phenomenon. A holistic approach is essential, integrating statistical analysis with a thorough evaluation of methodological rigor, potential biases, and alternative explanations. The replication crisis, where many positive results fail to be reproduced in subsequent studies, highlights the need for greater transparency and replication efforts in parapsychological research. The scientific community demands reproducible results before accepting a phenomenon as scientifically validated.

Another significant factor is the potential for cognitive biases to influence both the conduct and interpretation of parapsychological experiments. Confirmation bias, the tendency to favor information confirming pre-existing beliefs, can lead researchers to selectively focus on positive results and overlook or downplay negative findings. Expectation bias, where the experimenter's expectations influence the results, can also significantly distort the outcomes. To mitigate these biases, rigorous experimental designs are crucial, including the use of double-blind procedures, where neither the participant nor the experimenter knows the true experimental condition. The adoption of standardized protocols, detailed reporting procedures, and pre-registration of experiments can help to enhance the transparency and replicability of research findings. The use of independent observers, and the application of rigorous statistical methods to account for multiple comparisons, are essential steps towards minimizing the impact of bias.

Despite the challenges, controlled experiments continue to play a vital role in parapsychological research. While conclusive evidence for paranormal phenomena remains elusive, the pursuit of rigorous, replicable experiments is crucial for advancing our understanding of the nature of consciousness and the potential limits of human perception and interaction with the physical world. The future of parapsychological research hinges on a collaborative effort towards developing more robust experimental designs, employing advanced statistical techniques, and promoting a culture of openness and transparency within the scientific community. The quest to understand the boundaries of reality remains an ongoing journey, requiring both rigorous scientific investigation and an open mind to the possibilities that lie beyond current scientific paradigms. Only through persistent, rigorous experimentation and careful analysis can we hope to unravel the mysteries surrounding the seemingly paradoxical nature of paranormal phenomena.

The inherent challenges in studying purported paranormal phenomena are significantly amplified by the pervasive influence of bias and skepticism. While skepticism is a cornerstone of the scientific method, its extreme form can hinder objective investigation, prematurely dismissing potentially valuable research avenues. Conversely, confirmation bias, the tendency to favor information confirming pre-existing beliefs, can lead researchers to cherry-pick data, ignore contradictory evidence, and ultimately misrepresent findings. This delicate balance between healthy skepticism and open-mindedness is crucial in navigating the complexities of parapsychology.

One primary source of bias stems from the very nature of the phenomena under investigation. Many paranormal claims are inherently subjective and difficult to quantify. For example, reports of near-death experiences (NDEs) often involve profound emotional and spiritual transformations, making it challenging to establish objective criteria for verifying or falsifying the experiences. Similarly, claims of extrasensory perception (ESP) involve subtle mental processes that are difficult to isolate and measure using conventional scientific instruments. The subjective nature of these events inevitably leads to a higher degree of interpretive flexibility, allowing biases to creep in both during data collection and analysis.

The experimental design itself can inadvertently introduce biases. For instance, in experiments involving psychokinesis (PK), subtle flaws in the design of random number generators (RNGs) or the recording equipment could potentially influence the outcome, leading to spurious correlations. Researchers might unintentionally communicate their expectations to participants, consciously or unconsciously affecting their performance. This subtle form of experimenter bias can significantly influence the results, particularly in studies involving human subjects whose behavior can be easily affected by subtle cues.

The use of double-blind procedures, where neither the participant nor the experimenter knows the experimental condition, is a crucial step in mitigating experimenter bias. However, even with double-blind methods, subtle cues can inadvertently leak into the experiment. The researcher's body language, tone of voice, or even the arrangement of the experimental apparatus could unconsciously influence participant behavior. Therefore, meticulous attention to detail and rigorous control of all potential confounding variables are essential.

Another critical aspect is the selection of participants. Studies focusing on individuals who self-report strong paranormal abilities might introduce a sampling bias. These individuals might already possess a strong belief in the phenomenon being studied, which can increase the likelihood of positive results through expectation effects and confirmation bias. A more representative sample, encompassing individuals with diverse beliefs and experiences, is essential for producing unbiased results.

Data analysis presents another potential source of bias. The interpretation of statistically significant results often requires careful consideration of multiple comparison issues and the potential for Type I errors (false positives). Researchers might focus solely on statistically significant results while ignoring non-significant findings that might offer valuable insights into the complexities of the phenomenon. To mitigate this, a pre-registered analysis plan, specifying the statistical methods and criteria for determining significance before the data are collected, is highly recommended. This helps prevent researchers from selectively analyzing the data to achieve desired results.

Publication bias, the tendency for journals to publish positive rather than negative results, further exacerbates the problem. This leads to an overrepresentation of positive findings in the literature, creating a skewed perception of the overall state of knowledge on a given topic. The lack of publication of null or negative findings contributes to the difficulty in replicating results, hindering the advancement of the field. Initiatives to promote the publication of all study results, regardless of outcome, are essential for fostering a more balanced and comprehensive understanding.

The role of skepticism in parapsychological research is complex and multifaceted. While healthy skepticism is crucial for rigorous scientific investigation, it must not degenerate into outright dismissal of potential findings based solely on pre-conceived notions or the perceived impossibility of a phenomenon. Skepticism should be directed toward identifying methodological flaws, potential biases, and alternative explanations, rather than dismissing the phenomenon itself. A balanced approach involves critically examining both the evidence supporting and contradicting paranormal claims, while acknowledging the limitations of current scientific understanding. It is vital to distinguish between the healthy questioning of methodology and the categorical rejection of evidence without proper evaluation.

Moreover, the concept of "proof" itself warrants careful consideration. In science, it is rarely possible to achieve absolute proof; instead, we rely on accumulating evidence to strengthen or weaken a hypothesis. In parapsychology, the absence of conclusive proof does not automatically invalidate the possibility of paranormal phenomena. The fact that many purported paranormal events are rare, subtle, and difficult to replicate does not preclude their existence. The challenges in demonstrating these phenomena scientifically should not be interpreted as evidence against their existence but rather as a reflection of the limitations of current research methodologies.

A truly comprehensive approach necessitates a collaborative effort among researchers from diverse backgrounds, including physicists, psychologists, statisticians, and other relevant disciplines. Interdisciplinary collaborations can enrich the field by introducing diverse perspectives, methodologies, and analytical tools. This integrative approach can lead to more robust experimental designs, advanced statistical techniques, and more accurate interpretations of findings. Ultimately, the advancement of parapsychology, as with any scientific field, relies on a commitment to rigorous methodology, transparency, replication, and open-minded inquiry. The pursuit of understanding the limits of our current reality demands a blend of critical analysis and intellectual curiosity, constantly striving to refine our understanding and improve our methodologies. Only through such continuous refinement can we hope to unravel the enduring mysteries of the paranormal.

12

Exploring Consciousness Beyond the Physical

The mind-body problem, a philosophical conundrum that has plagued thinkers for millennia, lies at the heart of our attempts to understand consciousness and its potential relationship to realms beyond our immediate physical experience. The problem essentially boils down to this: how does the immaterial mind—our thoughts, feelings, consciousness—interact with the material body—our physical brain and nervous system? Is the mind simply a product of the brain's physical processes, an emergent property arising from complex neural activity, or is it something more, a separate entity interacting with the body in some mysterious way?

Materialism, a dominant perspective in contemporary science, posits that the mind is entirely dependent on the physical brain. According to this view, consciousness arises from the intricate network of neurons and synapses, and any alteration to the brain's structure or function directly impacts the mind. Damage to specific brain regions, for example, can lead to significant changes in personality, cognition, and even consciousness itself. Brain scans reveal correlations between mental activity and specific patterns of neural firing, lending further support to the materialistic perspective.

However, materialism struggles to fully explain certain phenomena. Subjective experiences, the "what it's like" aspect of consciousness, remain elusive. While we can measure brain activity correlated with a feeling of happiness or sadness, we can't directly translate that neural activity into the subjective experience itself. This "hard problem of consciousness," as philosopher David Chalmers termed it, highlights the limitations of purely materialistic explanations.

Dualism, an alternative perspective, proposes that the mind and body are distinct substances, interacting with each other but not reducible to one another. This view, championed by thinkers like René Descartes, suggests that the mind possesses a non-physical essence, perhaps existing in a separate realm. The mechanism of interaction between these two distinct substances remains a major challenge for dualism, leading to criticisms of its lack of explanatory power. The "interaction problem," as it's known, focuses on how a non-physical entity can causally affect a physical one. If the mind is truly separate from the brain, how can our thoughts influence our actions?

Idealism, another prominent philosophy, argues that reality is fundamentally mental, with the physical world existing only as a manifestation of consciousness. This radical view suggests that the mind is not merely separate from the body but is the ultimate ground of being, with the physical world being a projection of the mind itself. While this perspective offers a compelling explanation for the subjective nature of experience, it faces challenges in accounting for the apparent objectivity and consistency of the physical world. If reality is purely mental, how do we explain the shared experiences of multiple individuals, and the existence of physical laws independent of our minds?

Beyond these established philosophical viewpoints, recent scientific and parapsychological research suggests alternative models that could potentially bridge the gap between mind and body. Quantum physics, for instance, suggests that consciousness might play a more fundamental role in shaping reality than previously thought. The observer effect, where the act of observation influences the outcome of a quantum experiment, hints at a possible connection between consciousness and the physical world at a fundamental level. Some theories propose that consciousness might even be a fundamental aspect of the universe, existing alongside or even underpinning the physical laws we observe.

The exploration of near-death experiences (NDEs) also offers intriguing perspectives on the mind-body problem. NDEs frequently involve vivid out-of-body experiences, perceptions of leaving the physical body and observing events from a detached perspective. While skeptics often attribute

NDEs to physiological processes occurring during near-death situations, such as oxygen deprivation, the detailed and accurate accounts of some NDEs suggest that consciousness might persist even after the cessation of brain function. This raises questions about the relationship between consciousness and the brain, and suggests that consciousness might be more than just a product of neural activity.

Similarly, research into out-of-body experiences (OBEs) and other forms of altered states of consciousness offers further clues into the mind-body dynamic. While OBEs can be induced through various techniques, including sensory deprivation and hypnosis, some cases suggest a degree of conscious perception that transcends the limitations of the physical body. If individuals can perceive events from a location outside their physical body, this implies a dissociation between mind and body that challenges materialistic interpretations.

Furthermore, the study of psi phenomena, including telepathy, clairvoyance, and psychokinesis, poses significant challenges to materialistic models of mind. If individuals can influence physical systems or access information without relying on conventional sensory pathways, this suggests a level of interaction between mind and matter that exceeds the capabilities of our current understanding of the physical world. While rigorous scientific studies on psi phenomena are still lacking, their very existence, even if only as anecdotal evidence, warrants further exploration and consideration within the broader context of the mind-body problem.

The potential connection between consciousness and other realms also adds another layer of complexity to the mind-body problem. If consciousness is not limited to the physical body but can extend beyond it, perhaps even existing in other dimensions or planes of existence, then our understanding of reality needs to expand beyond purely materialistic explanations. Some theoretical models propose that variations in density might create different planes of existence, potentially explaining phenomena like ghosts or otherworldly encounters. If these alternative realms exist, consciousness could bridge the gap between them, providing a pathway for interaction between different levels of reality.

In conclusion, the mind-body problem remains a central challenge in our pursuit of understanding consciousness and its potential connection to other realms. While materialism offers a compelling framework for understanding many aspects of mind and brain function, it falls short of explaining subjective experience and certain phenomena that suggest a more profound connection between mind and matter, or even between mind and other dimensions. Dualism, idealism, and other philosophical perspectives offer alternative viewpoints, but each faces its own challenges and limitations. The integration of insights from parapsychology, quantum physics, and the study of altered states of consciousness could potentially lead to a more comprehensive and nuanced understanding of the mind-body problem, ultimately expanding our vision of reality beyond our current physical confines.

The search for a unified theory of consciousness demands open-mindedness, rigorous scientific inquiry, and a willingness to explore unconventional perspectives, leading us further down the path of understanding the profound mysteries of existence. The current limitations in our understanding should not be equated with the impossibility of a more unified model. Rather, they highlight the need for continued investigation and a broader perspective that accommodates diverse viewpoints and approaches. The journey towards a complete comprehension of consciousness and its implications is a challenging but ultimately rewarding pursuit.

The exploration of consciousness beyond the physical inevitably leads us to consider the intriguing possibility of its connection to quantum phenomena, specifically quantum entanglement. Entanglement, a bizarre yet well-established principle of quantum mechanics, describes a scenario where two or more particles become linked in such a way that they share the same fate, regardless of the distance separating them. Measuring the properties of one entangled particle instantaneously reveals the properties of its partner, even if they are light-years apart. This seemingly instantaneous correlation defies classical physics and challenges our understanding of locality and causality.

The implications of entanglement for our understanding of consciousness are profound. Some physicists and parapsychologists propose that consciousness might be intricately woven into the fabric of quantum reality, potentially playing a role in the collapse of the wave function—the process by which a quantum system transitions from a superposition of states to a single, definite state upon measurement. The observer effect, a well-documented phenomenon in quantum physics, suggests that the act of observation itself influences the outcome of a quantum experiment. This raises the question: could consciousness be the "observer" that collapses the wave function, potentially influencing the behavior of entangled particles?

The "many-worlds interpretation" of quantum mechanics, while controversial, provides a framework for exploring this connection. This interpretation suggests that every quantum measurement causes the universe to split into multiple branches, each representing a different possible outcome. In this framework, consciousness might not collapse the wave function but rather selectively experience one particular branch of reality, while other branches continue to exist independently. This perspective suggests a much more dynamic and participatory role for consciousness in shaping reality than traditional interpretations of quantum mechanics allow.

However, the precise mechanism by which consciousness might influence or be influenced by quantum events remains a mystery. Some theories propose that consciousness operates at a quantum level, interacting directly with entangled particles through yet-unknown mechanisms. This could potentially explain phenomena such as telepathy or clairvoyance, which involve the transfer of information between minds without the use of conventional sensory pathways. If consciousness can influence quantum systems, then the apparent non-locality of entanglement could provide a mechanism for instantaneous communication between minds, irrespective of physical distance.

Other models suggest that the relationship between consciousness and entanglement is more indirect. Perhaps the brain, with its vast network of neurons and synapses, acts as a highly sensitive quantum system, capable of detecting and responding to subtle quantum fluctuations. This could allow the brain to access and process information from entangled particles, providing a subtle yet significant influence on conscious experience. For instance, certain neural processes might be particularly sensitive to quantum entanglement, potentially explaining the subtle correlations observed in some studies of psychic phenomena.

The challenge in exploring the connection between consciousness and quantum entanglement lies in the difficulty of designing and conducting rigorous experiments. Quantum systems are notoriously fragile, and the precise nature of the interaction between consciousness and quantum events remains elusive. Moreover, the subjective nature of consciousness makes it difficult to quantify and measure objectively. However, ongoing research in quantum biology, which explores the role of quantum mechanics in biological systems, offers promising avenues for investigating this connection.

Several promising areas of investigation are currently under exploration. One approach involves studying the effects of focused attention and intention on quantum systems. If consciousness can influence the collapse of the wave function, then it might be possible to detect subtle changes in the behavior of entangled particles when observers focus their attention on them. Such experiments require sophisticated equipment and careful control of experimental conditions, but they offer the potential to provide empirical evidence for a connection between consciousness and quantum phenomena.

Another avenue of research involves studying the brain's quantum properties. Some scientists suggest that certain aspects of brain function, such as memory and consciousness, might rely on quantum coherence—a phenomenon where quantum systems maintain a coherent superposition of states. If this hypothesis holds true, then the brain could act as a natural quantum computer, capable of processing information in ways that are impossible for classical computers. This would have profound implications for our understanding of consciousness and its potential relationship to quantum entanglement.

The exploration of near-death experiences (NDEs) also lends itself to consideration within this quantum framework. The vivid and often detailed accounts of out-of-body experiences and perceptions of events from a detached perspective reported in NDEs could potentially be explained by a dissociation of consciousness from the physical brain, allowing consciousness to interact with quantum realms beyond the constraints of the physical body. If consciousness is not strictly confined to the physical brain, then its ability to influence or be influenced by quantum phenomena might be enhanced.

Furthermore, the study of altered states of consciousness, such as those induced through meditation, hypnosis, or psychedelic drugs, offers insights into the potential connections between consciousness, quantum processes, and other realms. These states of consciousness are characterized by altered perceptions, heightened awareness, and sometimes even experiences that transcend the boundaries of space and time. The investigation of these experiences from a quantum perspective could reveal novel connections between consciousness, quantum mechanics, and the nature of reality itself.

It's crucial to acknowledge the skepticism surrounding the connection between consciousness and quantum entanglement. Many scientists remain unconvinced by the evidence presented so far, citing the lack of rigorous experimental data and the difficulty of separating subjective experiences from physical processes. However, the theoretical possibility, supported by the inherent strangeness of quantum mechanics and the persistent mysteries surrounding consciousness, remains a compelling avenue for scientific inquiry.

The exploration of this connection requires a multidisciplinary approach, drawing on insights from quantum physics, neuroscience, parapsychology, and philosophy of mind. It demands open-mindedness, rigorous scientific inquiry, and a willingness to challenge existing paradigms. The potential implications of a unified theory of consciousness encompassing quantum entanglement are vast, extending beyond the realm of theoretical physics to encompass our understanding of reality, the nature of existence, and the potential for human consciousness to interact with the universe in unexpected and profound ways. The ongoing research in this area pushes the boundaries of our understanding, potentially unveiling a deeper connection between the subjective and objective realms, a bridge between the inner world of experience and the outer world of physical phenomena. The journey, while fraught with challenges and uncertainties, promises discoveries that could revolutionize our understanding of the universe and our place within it. The persistent pursuit of this knowledge, irrespective of the challenges, underlines the importance of exploring these unconventional avenues, always striving for rigorous methodology and a commitment to uncovering the truth, regardless of how strange or unexpected it may be.

Near-death experiences (NDEs) represent a compelling area of study when exploring consciousness beyond the physical. These profound events, often involving a sense of detachment from the body, visions of light and deceased loved ones, and a review of one's life, challenge our conventional understanding of death and the limitations of the physical brain.

While often dismissed as hallucinations or oxygen deprivation-induced phenomena, the striking similarities across diverse cultures and individuals warrant a closer examination, particularly within the context of our exploration of multi-dimensional realities and the potential for consciousness to transcend physical constraints.

One of the most intriguing aspects of NDEs is the recurring theme of out-of-body experiences (OBEs). Many individuals report floating above their bodies, observing medical procedures being performed on them from a detached perspective. These accounts are often remarkably accurate, detailing events that the individual could not have known through ordinary sensory perception. Some researchers suggest these OBEs might indicate a temporary separation of consciousness from the physical body, allowing consciousness to interact with the environment in a non-physical state. This aligns with our exploration of density variations; perhaps consciousness, freed from the constraints of dense physical matter, can exist and perceive in a less dense realm, accessing information inaccessible to the physically embodied self.

Furthermore, the common experience of encountering a bright light, often described as overwhelmingly peaceful and loving, is noteworthy. This "light" could be interpreted in several ways within our proposed framework. It might represent a shift in consciousness to a higher-density plane of existence, where light is not simply electromagnetic radiation but a fundamental aspect of the realm's structure. Alternatively, it could be a symbolic representation of the transition between different states of consciousness, a metaphorical gateway between the physical and non-physical realms. The consistency of this experience across different cultures and individuals suggests a potential underlying reality, rather than a purely subjective hallucination.

The reports of encountering deceased loved ones further deepen the mystery. These encounters often involve emotional reunions, offering comfort and reassurance. If we postulate the existence of other dimensions or realms with varying densities, these experiences might reflect interactions with consciousnesses inhabiting those planes of existence, potentially communicating through non-physical channels. The detailed and emotionally resonant nature of these encounters suggests a significant interaction, challenging the materialistic view that consciousness ceases to exist upon death. This aligns with the concept of consciousness as a fundamental aspect of the universe, existing independently of the physical body.

Another prominent feature of NDEs is the "life review," where individuals report a rapid and comprehensive recollection of their past lives, encompassing both positive and negative experiences. This review often leads to profound self-reflection and a shift in values and priorities. Within our framework, this could be interpreted as a process of integrating experiences across different dimensions or density levels. The life review might represent a consolidation of consciousness's experiences within different realms, leading to a more holistic and integrated self-understanding. This integration might explain the reported changes in personality and perspective following an NDE, reflecting a transformation in consciousness.

The clinical setting in which many NDEs occur also offers valuable insights. Individuals in cardiac arrest, where oxygen supply to the brain is severely reduced, often report vivid NDEs. However, the complexity and coherence of these experiences seem at odds with the assumption that they are merely hallucinations resulting from oxygen deprivation. The detailed accounts and accurate observations suggest a more structured and organized mental process than would be expected during extreme physiological distress. This raises the possibility that consciousness might be able to function independently of the brain's normal oxygen-dependent processes, operating through a less materially constrained mechanism.

Furthermore, the universality of NDE themes across different cultures and religions provides compelling evidence for a potentially universal underlying experience. Regardless of cultural background or belief system, the core elements of NDEs—the out-of-body experience, the encounter with light, the life review—remain remarkably consistent. This consistency suggests that these experiences tap into a fundamental aspect of human consciousness, transcending the boundaries of culture and belief.

However, a significant challenge remains in verifying the claims made during NDEs. The subjective nature of the experience makes it difficult to establish objective evidence. Verifying the accuracy of out-of-body observations, for instance, requires meticulous investigation and corroboration, which isn't always possible. Moreover, the potential for misinterpretation, memory biases, and suggestion cannot be completely ruled out. Nonetheless, the consistent and detailed nature of many NDE accounts demands a careful and open-minded investigation.

While rigorous scientific research on NDEs is still in its early stages, several avenues are currently being explored. Researchers are studying brain activity during near-death situations, attempting to correlate physiological changes with the reported experiences. Advances in neuroimaging technologies may shed light on the neural mechanisms involved in NDEs, potentially revealing whether consciousness can exist independently of typical brain function. Cross-cultural studies are also underway, aiming to identify universal patterns and potential differences in NDE accounts across diverse cultures and belief systems.

The challenge lies in developing methodologies that can adequately address the subjective nature of the experiences while maintaining scientific rigor. This requires a multidisciplinary approach, bringing together expertise from neuroscience, psychology, parapsychology, and philosophy to explore the complexities of NDEs from various perspectives.

Technological advancements in brain imaging and data analysis will also play a critical role in uncovering potential neural correlates of these extraordinary experiences.

Considering NDEs within the context of our exploration of density, dimensions, and consciousness outside of the physical body suggests the possibility that NDEs represent glimpses into other planes of existence, perhaps temporarily accessible due to the extreme physiological conditions. This aligns with the model where variations in density create different realms accessible to consciousness under specific circumstances. Further investigation into these experiences might reveal more about the nature of consciousness and its interaction with the universe beyond our current understanding of physics and biology. The study of NDEs offers a unique window into the possible existence of realms beyond our ordinary perception, prompting a deeper inquiry into the fundamental nature of reality and the boundless potential of human consciousness. The more we investigate, the closer we may come to understanding the true nature of reality and our place within it. The journey remains long, but the potential rewards of uncovering the secrets held within these experiences are vast and compelling, pushing the boundaries of science and philosophy alike.

Out-of-body experiences (OBEs) represent a particularly intriguing facet of consciousness studies, often intertwined with near-death experiences but possessing a distinct character. While NDEs frequently involve a sense of detachment from the physical body, OBEs often focus on the reported sensation of leaving the body entirely and perceiving the environment from an external perspective. The scientific community remains divided on the nature and interpretation of OBEs, with explanations ranging from neurological malfunctions to psychological phenomena. However, the consistent features reported across numerous accounts warrant a deeper investigation, especially within the framework of exploring dimensions beyond our ordinary perception.

One of the central challenges in studying OBEs lies in the subjective nature of the experiences. Verifying the accuracy of perceptions reported during an OBE is extremely difficult. While some individuals report observing events that could later be verified, this verification is often circumstantial and susceptible to various biases. The possibility of memory distortion, suggestibility, and even conscious or unconscious fabrication cannot be entirely dismissed. The lack of objective, measurable data significantly hampers the scientific investigation. However, the sheer volume of consistent reports from individuals across different cultures and backgrounds suggests a phenomenon deserving of rigorous scientific scrutiny, even if the precise mechanisms remain elusive.

Neuroscientific perspectives offer potential explanations for OBEs, primarily focusing on brain activity and sensory processing. Some researchers suggest that OBEs might arise from disruptions in the brain's sensory integration processes, leading to a disconnection between the perceived body image and the actual sensory input. Neurological conditions affecting the parietal lobe, responsible for spatial awareness and body perception, have been implicated in OBE-like experiences. Damage or dysfunction in this region can lead to distortions in body image and the feeling of detachment from one's body. Similarly, temporal lobe epilepsy has been associated with experiences resembling OBEs, suggesting that aberrant electrical activity in specific brain regions might play a crucial role.

Furthermore, the influence of sleep and dreaming on OBEs cannot be ignored. Many OBEs occur during sleep paralysis, a state where individuals are conscious but unable to move their bodies. The vivid imagery and sensory hallucinations that can occur during this state are often remarkably similar to OBE accounts. The overlap between sleep paralysis and OBEs supports the idea that neurological processes related to sleep and dreaming might contribute to the experience. However, this does not negate the possibility that OBEs might represent something beyond mere neurological dysfunction. The level of detail and accuracy reported in some OBEs, particularly concerning events that the individual could not have otherwise known, pose a significant challenge to these exclusively neurobiological interpretations.

Psychological factors also appear to play a significant role in OBEs. Stress, anxiety, and trauma can trigger altered states of consciousness that might include OBEs. Furthermore, certain psychological predispositions, such as a strong imagination or a tendency towards dissociation, might increase the likelihood of experiencing OBEs. The impact of cultural beliefs and expectations cannot be overlooked. Societal narratives and expectations surrounding OBEs might influence both the likelihood of experiencing them and the way in which the experiences are reported and interpreted. A comprehensive understanding of OBEs requires a multi-faceted approach, integrating insights from both neuroscience and psychology.

A potential connection between OBEs and our model of multi-dimensional realities and varying densities opens up fascinating avenues of exploration. If consciousness is not solely confined to the physical brain, but rather exists as a fundamental aspect of the universe, it is conceivable that under certain conditions, consciousness might temporarily detach from the dense physical body and interact with other dimensions or density levels. OBEs could represent instances of this temporary separation, allowing consciousness to perceive the environment from a non-physical perspective. This perspective might afford access to information and experiences unavailable to the physically embodied self. The accuracy of certain OBE accounts, particularly those reporting details verifiable later, raises the possibility of accessing information through channels beyond our typical sensory perception.

This concept aligns with the idea of consciousness as an informational field existing beyond the confines of the physical body. In this model, OBEs are not merely hallucinations or neurological artifacts but rather instances of consciousness interacting with the universe in a way currently poorly understood by conventional physics. The variations in density, as previously discussed, could create different realms or planes of existence, each with its own unique characteristics and accessibility to consciousness. OBEs might represent momentary shifts in consciousness from a denser physical realm to a less dense one, temporarily granting access to information and perceptions unavailable in the physical state.

Exploring this concept necessitates a paradigm shift, moving away from purely materialistic explanations toward a more holistic view of consciousness. It challenges the traditional notion that consciousness is entirely dependent on the brain's biological processes. Instead, it proposes that consciousness might be a fundamental aspect of the universe, capable of existing independently of physical constraints. This perspective opens up numerous avenues for future research, including the investigation of potential correlations between OBE characteristics and variations in brainwave activity, electromagnetic fields, or other physical parameters.

Moreover, the role of intentionality in OBEs remains largely unexplored. Some researchers suggest that the intention to leave one's body might play a crucial role in initiating an OBE. If consciousness can indeed interact with other dimensions or density levels, the conscious intention to shift to another realm might increase the likelihood of experiencing an OBE. This opens up further avenues of exploration, potentially involving controlled studies investigating the impact of intention on the occurrence and characteristics of OBEs.

In conclusion, while many questions surrounding OBEs remain unanswered, their consistent reporting across cultures and the challenging aspects of some experiences necessitate further, rigorous investigation. A multidisciplinary approach, combining insights from neuroscience, psychology, and a broader exploration of the nature of consciousness, is crucial. By integrating these diverse perspectives, we might move closer to a more comprehensive understanding of OBEs, possibly gaining profound insights into the nature of consciousness and the interconnectedness of our perceived reality with potentially unseen dimensions or realms. The scientific exploration of OBEs represents a significant step towards unraveling the mysteries of consciousness and potentially revealing aspects of reality far beyond our current understanding. The journey requires patience, methodological rigor, and a willingness to explore beyond the confines of conventional scientific paradigms.

The prevailing materialistic worldview, deeply entrenched in modern science, posits that consciousness is solely a product of physical processes within the brain. This perspective, while offering valuable insights into the neurological correlates of consciousness, struggles to fully account for a range of experiences that appear to transcend the purely physical. Phenomena such as near-death experiences (NDEs), out-of-body experiences (OBEs), and various forms of psi phenomena – including telepathy, clairvoyance, and precognition – challenge the completeness of materialism. These experiences, often reported with striking consistency across cultures and historical periods, suggest the existence of aspects of consciousness that cannot be readily explained by purely physical mechanisms.

Materialism, in its strictest form, asserts that the physical world is all that exists, and that all phenomena, including consciousness, can ultimately be reduced to physical interactions. While this perspective has been remarkably successful in explaining many aspects of the natural world, its limitations become apparent when confronted with the subjective nature of consciousness. The "hard problem of consciousness," as famously articulated by philosopher David Chalmers, highlights the difficulty of explaining how physical processes give rise to subjective experience – the "what it's like" aspect of consciousness. Materialism, in its current form, offers no satisfactory answer to this fundamental question.

One of the primary challenges faced by materialistic accounts of consciousness is the lack of a clear mechanism for bridging the gap between objective physical processes and subjective experience. Neuroscience has made tremendous progress in mapping brain activity and correlating specific patterns of neuronal firing with different cognitive functions. However, this correlation does not necessarily imply causation. Simply identifying neural correlates of consciousness does not explain how these physical processes actually *produce* subjective experience. The subjective feeling of redness, for example, cannot be reduced to a specific pattern of neuronal activity, despite the existence of a strong correlation between the two.

Furthermore, the materialistic paradigm struggles to accommodate the seemingly non-local aspects of consciousness suggested by various parapsychological phenomena. Telepathy, for example, implies a form of communication that transcends the limitations of space and time, suggesting a non-physical interaction between minds. Similarly, precognitive experiences, where individuals report experiencing events before they occur, challenge the strictly linear, cause-and-effect model of time embraced by materialism. While skeptical interpretations of these phenomena often invoke coincidence, statistical anomalies, or flaws in experimental design, the sheer volume and consistency of such reports, across various cultures and historical periods, warrant serious consideration.

Alternative models of consciousness, moving beyond the strictures of materialism, offer potential solutions to these challenges. One such approach is panpsychism, which proposes that consciousness is a fundamental property of the universe, present in all matter, albeit to varying degrees. In this view, consciousness is not a mere epiphenomenon of complex physical systems but rather an inherent aspect of reality itself. This perspective offers a potential resolution to the hard problem of consciousness by suggesting that subjective experience is not an emergent property of complex systems but a fundamental property of the universe.

Another alternative is the "information field" theory, which posits that consciousness is associated with an informational field that extends beyond the confines of the physical brain. This field might interact with the physical world, influencing and being influenced by physical processes, but it is not itself reducible to physical processes. This model offers a potential explanation for phenomena like OBEs and NDEs, where individuals report experiencing consciousness outside of their physical bodies. The informational field could be understood as a form of non-local consciousness, capable of interacting with the physical world in ways that are not readily explained by conventional physics.

The concept of a "multiverse" or "many-worlds" interpretation of quantum mechanics also provides a framework for understanding consciousness beyond the purely material. In this interpretation, multiple universes exist simultaneously, each representing a different possible outcome of quantum events. Consciousness, in this framework, might be understood as a fundamental aspect of these multiple universes, capable of interacting with and influencing them. This perspective suggests that our perceived reality is merely one among many, and that our consciousness might be capable of interacting with other realities beyond our ordinary perception.

Furthermore, recent advancements in quantum physics suggest a deeper connection between consciousness and the physical world than previously thought. The observer effect in quantum mechanics, where the act of observation influences the outcome of a quantum event, implies a fundamental role of consciousness in the structure of reality. This raises the possibility that consciousness is not simply a passive observer of the universe but an active participant in shaping it.

The exploration of consciousness beyond the limits of materialism requires a multidisciplinary approach, integrating insights from neuroscience, psychology, physics, and parapsychology. It necessitates a willingness to move beyond established paradigms and embrace unconventional explanations. The challenges are significant, but the potential rewards are equally immense. By acknowledging the limitations of a purely materialistic worldview and embracing the possibility of non-physical aspects of consciousness, we can unlock a deeper understanding of the nature of reality and our place within it. This journey demands intellectual humility, open-mindedness, and a commitment to exploring the full spectrum of human experience. Only through such a comprehensive approach can we hope to unravel the mysteries of consciousness and gain a more complete picture of our place in the cosmos.

13

The Interconnectedness of All Things

The exploration of consciousness beyond the confines of materialistic reductionism naturally leads us to consider more radical perspectives on the nature of reality itself. One such perspective, gaining increasing traction within both physics and parapsychology, is the holographic universe hypothesis. This hypothesis, far from being mere science fiction, rests upon significant theoretical foundations and offers a compelling framework for understanding the interconnectedness of all things, including the apparent non-local aspects of consciousness.

The holographic principle, initially conceived within the context of string theory and black hole thermodynamics, suggests that the information describing a volume of space can be encoded on its boundary. Imagine a three-dimensional object projected onto a two-dimensional surface, like a hologram. The two-dimensional image contains all the information necessary to reconstruct the three-dimensional object. Similarly, the holographic principle proposes that our three-dimensional universe, with all its complexity and apparent solidity, might be a projection from a higher-dimensional space. This higher-dimensional space, often referred to as the "holographic screen," contains all the information necessary to describe our universe.

This idea has profound implications for our understanding of space, time, and causality. If our universe is a hologram, then the seemingly separate events and objects within it are fundamentally interconnected, their interactions determined by the underlying information encoded on the holographic screen. This could provide a potential explanation for phenomena like entanglement, where two quantum particles remain linked regardless of the distance separating them. Their interconnectedness is not a result of physical interaction across space but a reflection of their shared information on the holographic screen.

The holographic principle also challenges our conventional understanding of locality. In a holographic universe, information is not localized within specific points in space but is distributed across the entire holographic screen. This non-local nature of information could help explain certain parapsychological phenomena, such as telepathy or precognition. If consciousness is somehow linked to the information on the holographic screen, then instantaneous communication between minds or access to future events could be possible.

Consider the implications for our understanding of time. If our universe is a projection, the concept of a linear flow of time could become less rigid. The holographic screen might contain information representing all moments of our universe's existence simultaneously. This would suggest that the past, present, and future are not distinct but rather different aspects of a unified whole. Precognitive experiences could then be interpreted as a glimpse of information already present on the holographic screen, rather than a violation of causality.

The holographic principle is not without its challenges. While it is mathematically consistent within certain theoretical frameworks, a complete and comprehensive model of a holographic universe remains elusive. Several key questions remain unanswered. What is the nature of the holographic screen? What are the physical mechanisms by which information is encoded and projected? How does consciousness relate to this information? These are fundamental questions that require further investigation, drawing upon insights from both theoretical physics and the ongoing research in parapsychology.

The exploration of the holographic universe hypothesis has profound implications for our understanding of consciousness. If our universe is fundamentally interconnected, as the holographic principle suggests, then this could provide a basis for understanding the non-local aspects of consciousness observed in phenomena like telepathy and remote viewing. Consciousness, in this framework, might not be confined to the physical brain but could be distributed across the holographic screen, enabling non-local interactions.

The connection between the holographic universe hypothesis and parapsychology is not simply speculative. Many researchers in the field of parapsychology have found the holographic principle to be a valuable framework for understanding seemingly inexplicable phenomena. The non-local nature of information, as described by the holographic principle, could provide a theoretical underpinning for the apparent non-locality of consciousness observed in various psi phenomena. For instance, telepathic communication could be interpreted as an interaction between conscious entities through the shared information on the holographic screen, rather than through any conventional physical mechanism.

The implications extend further. If our consciousness is not limited to our physical brains but is part of a larger, interconnected system, then our understanding of death and mortality may need to be revisited. The information that constitutes our consciousness may not be destroyed upon physical death but rather may continue to exist on the holographic screen, potentially in an altered form. This could offer a scientific framework for the persistent reports of near-death experiences and other afterlife phenomena.

The holographic universe hypothesis also offers a framework for understanding the apparent fine-tuning of the universe for life. The precise values of physical constants that allow for the existence of life could be seen as a reflection of the underlying information encoded on the holographic screen. This information, in a sense, "selects" the parameters that lead to a universe capable of supporting life, suggesting a deeper level of purpose and meaning within the cosmos.

However, it's crucial to acknowledge that the holographic universe hypothesis remains a theoretical framework. While it offers a compelling and potentially unifying perspective, experimental verification remains a significant challenge. The scales involved are far beyond our current technological capabilities, requiring a fundamental shift in our understanding of physics and our methods of investigation. Nevertheless, the theoretical implications are so profound and the potential for explaining a range of seemingly disparate phenomena so great that continued exploration and investigation are warranted.

The integration of this hypothesis with the findings of parapsychology suggests a fascinating and potentially revolutionary shift in our worldview. It could lead to a more holistic and interconnected understanding of reality, where consciousness plays an active, not merely passive, role in shaping the universe. This shift in perspective could profoundly impact not only our scientific understanding but also our philosophical and spiritual outlook. The journey of exploration is far from over, yet the potential rewards – a deeper understanding of consciousness, reality, and our place within the cosmos – are immeasurable. The quest to unravel these mysteries demands continued rigorous scientific investigation, coupled with an open mind and a willingness to explore the seemingly impossible.

The interconnectedness of all things, hinted at by the holographic universe hypothesis, could revolutionize our understanding of the universe, and potentially, our place within it. This is a journey that demands our collective effort, a symphony of scientific inquiry and philosophical reflection, all playing in harmony to uncover the profound secrets of existence. The pursuit of this understanding is a venture that transcends the purely scientific; it's a quest for meaning, for connection, and for a deeper understanding of our place in the grand cosmic tapestry.

Quantum entanglement, a phenomenon predicted by quantum mechanics and experimentally verified countless times, presents a profound challenge to our classical understanding of reality. It describes a situation where two or more quantum particles become linked in such a way that they share the same fate, regardless of the distance separating them. Measuring the property of one particle instantaneously determines the corresponding property of the other, even if they are light-years apart. This instantaneous correlation defies the limitations imposed by the speed of light, a cornerstone of classical physics. Einstein famously referred to this as "spooky action at a distance," highlighting the counterintuitive nature of entanglement.

The non-locality inherent in quantum entanglement implies that the entangled particles are not truly independent entities but are instead interconnected in a fundamental way. Their properties are not pre-determined and localized within each particle, but rather emerge from a shared, non-local reality. This suggests a deeper interconnectedness within the universe than classical physics allows. Instead of a universe composed of isolated, independent objects interacting through local forces, quantum entanglement suggests a universe where objects are inherently linked, regardless of spatial separation.

This non-locality raises profound questions about the nature of space and time. If information can be transmitted instantaneously across vast distances through entanglement, it challenges the classical notion of spacetime as a fixed and immutable framework. It suggests that spacetime might be more malleable, or even emergent, rather than a fundamental component of reality. The very fabric of space and time could be intertwined with the quantum correlations between entangled particles. Imagine a universe where spatial separation is not an absolute barrier, where information can transcend the limitations of distance.

The implications of entanglement extend beyond the purely physical realm. The apparent instantaneous correlation between entangled particles suggests a form of non-local connection that bears a striking resemblance to certain phenomena reported in parapsychology, such as telepathy or remote viewing. While the mechanisms are certainly different, the non-local nature of both quantum entanglement and these parapsychological phenomena suggests a fundamental interconnectedness that transcends our everyday experience of space and time.

Consider the implications for our understanding of consciousness. If the universe is fundamentally non-local, as quantum entanglement suggests, it opens up the possibility that consciousness itself might be non-local. Our subjective experience of being localized within our physical bodies might be an illusion, a projection of a deeper, more interconnected consciousness that transcends the boundaries of space and time. The entangled particles, existing in a state of interconnectedness, might serve as a rudimentary analogy for the potential interconnectedness of consciousnesses.

The non-locality of quantum entanglement is not simply a quirk of quantum mechanics; it has significant implications for our understanding of the universe and our place within it. It challenges the classical worldview, demanding a re-evaluation of fundamental concepts like space, time, causality, and even consciousness. The instantaneous correlation between entangled particles suggests a deeper level of interconnectedness than we have ever imagined, potentially offering a new perspective on the seemingly paradoxical phenomena studied in parapsychology.

The challenge, however, lies in bridging the gap between the microscopic realm of quantum entanglement and the macroscopic realm of human experience. While we have ample evidence for entanglement at the quantum level, the mechanisms by which this non-locality could manifest at the macroscopic level, particularly in relation to consciousness and parapsychological phenomena, remain largely unknown. This requires further research, integrating insights from both physics and parapsychology, to explore the potential connections between these seemingly disparate domains.

One promising avenue of investigation involves exploring the role of quantum coherence in biological systems. While the details are still largely unknown, there is growing evidence that quantum effects might play a crucial role in various biological processes, including photosynthesis, bird navigation, and even cognitive functions. If quantum coherence extends to macroscopic scales, it could provide a mechanism for non-local interactions within biological systems, potentially explaining some of the phenomena observed in parapsychology.

Another avenue of research involves exploring the relationship between quantum entanglement and the structure of spacetime itself. Several theoretical models suggest that spacetime might be an emergent phenomenon, arising from the underlying quantum correlations between fundamental particles. If this is the case, it could provide a theoretical basis for the non-local interactions observed in quantum entanglement and, potentially, in parapsychological phenomena. This line of inquiry necessitates a deeper understanding of quantum gravity, a field that seeks to unify general relativity with quantum mechanics.

The exploration of quantum entanglement and its implications for our understanding of reality requires a paradigm shift, moving away from a purely materialistic view of the universe to a more holistic and interconnected perspective. This requires a willingness to consider the possibility of non-local interactions, not only at the quantum level but also at the macroscopic level, including the potential involvement of consciousness.

The concept of non-locality challenges the classical notion of causality, where effects follow causes in a linear and deterministic way. In a non-local universe, cause and effect might be less clearly defined, with events seemingly connected in a way that transcends spatial and temporal separation. This non-linearity could potentially explain phenomena like precognition, where future events appear to influence present experience.

Moreover, the non-locality inherent in quantum entanglement challenges the notion of objective reality. The act of measurement plays a fundamental role in quantum mechanics, influencing the state of the system being observed. This suggests that the observer is not merely a passive spectator but an active participant in shaping reality. The very act of observing an entangled particle instantaneously affects the state of its entangled partner, regardless of the distance between them. This suggests a deeper interplay between consciousness and reality than traditionally assumed.

The study of quantum entanglement, therefore, is not simply an exercise in theoretical physics. It has profound implications for our understanding of consciousness, reality, and our place within the universe. It challenges our fundamental assumptions about space, time, and causality, paving the way for a more holistic and interconnected understanding of existence. The implications are far-reaching, extending from the foundations of physics to the most profound questions about consciousness and the nature of reality itself. The ongoing research in this field holds the potential to revolutionize not only our scientific understanding but also our philosophical and spiritual perspectives on the cosmos and our place within it. The journey of exploration is far from over, but the potential rewards – a deeper understanding of the universe and ourselves – are truly immeasurable. The pursuit of this understanding is a journey of both scientific discovery and philosophical reflection, requiring a collaborative effort that bridges the gap between scientific rigor and open-minded exploration of the unknown. This interconnectedness, hinted at by the subtle dance of entangled particles, invites us to reconsider our understanding of reality, pushing the boundaries of scientific thought and inviting a more comprehensive view of the universe and our role within its grand design. The implications stretch beyond mere scientific curiosity; they touch upon the very essence of existence, challenging us to redefine our perception of reality and our place within the vast cosmic tapestry.

The exploration of quantum entanglement has led us to the precipice of a profound realization: the universe may be far more interconnected than our classical worldview allows. This interconnectedness, extending beyond the realm of physics into the speculative but intriguing domain of parapsychology, hints at the existence of a universal consciousness, a cosmic mind if you will, that underpins the fabric of reality. One concept that resonates deeply with this idea is that of the Akashic Records.

The Akashic Records, a term originating in ancient Sanskrit, are often described as a universal database or library containing the complete history of every soul and every event that has ever occurred, is occurring, or will occur. This isn't a physical library, of course, but rather a metaphysical repository, a kind of cosmic memory containing all information and experiences within the universe. This concept, while lacking direct scientific evidence, aligns with the increasingly interconnected view of reality emerging from quantum physics.

If the universe is indeed fundamentally non-local, as suggested by quantum entanglement, the existence of a universal, interconnected consciousness becomes less improbable. The instantaneous correlation between entangled particles suggests a level of interconnectedness that transcends spatial and temporal limitations. Information, seemingly, isn't bound by the speed of light, suggesting a deeper reality underpinning the physical world. Could the Akashic Records be a manifestation of this deeper reality, a cosmic repository of information existing outside of spacetime as we understand it?

The concept of the Akashic Records bears a resemblance to certain theoretical models in physics, such as the holographic universe hypothesis. This hypothesis proposes that the universe is fundamentally a two-dimensional information structure projected onto a three-dimensional space-time. The entirety of the universe's information is encoded on this two-dimensional surface, much like a hologram. In this view, every point in the universe contains the complete information of the entire universe. The Akashic Records, if they exist, could be interpreted as a manifestation of this underlying information structure, accessible, perhaps, only through altered states of consciousness.

Furthermore, the concept of a universal consciousness resonates with the findings of quantum mechanics regarding the role of the observer in shaping reality. The act of observation, as we've seen, profoundly influences the quantum system being observed. This suggests an interconnectedness between the observer and the observed, a blurring of the lines between subject and object. A universal consciousness could be seen as the ultimate observer, shaping and interacting with the universe in a fundamental way.

The challenge in exploring these ideas lies in bridging the gap between the theoretical framework of physics and the metaphysical nature of the Akashic Records. Scientific methods, designed to study the physical world, struggle to grapple with concepts that exist outside the realm of empirical observation. Nonetheless, the potential link between quantum entanglement, the holographic universe hypothesis, and the concept of a universal consciousness warrants further investigation.

Consider the implications of such a system. If the Akashic Records hold the complete history of everything, it implies a profound interconnectedness between all events and all beings. Every action, every thought, every emotion resonates throughout the universe, leaving its imprint in the cosmic memory. This concept challenges our traditional linear view of time and causality, suggesting a more holistic and interconnected reality where everything is intricately linked.

Moreover, the accessibility of the Akashic Records, often described as being accessible through altered states of consciousness such as deep meditation or lucid dreaming, suggests a deeper connection between consciousness and the universe itself. The possibility of accessing this cosmic memory opens up profound implications for our understanding of ourselves and our place within the universe. It suggests a level of potential for personal growth and understanding far beyond our current conceptions.

The exploration of the Akashic Records and universal consciousness necessitates a shift in perspective. We need to move beyond purely materialistic interpretations of reality and embrace a more holistic, interconnected worldview that incorporates the insights of both science and metaphysics. This involves a willingness to explore unconventional ideas, challenging our preconceived notions and embracing the potential for deeper levels of reality. Scientific rigor should not preclude the exploration of metaphysical concepts, but rather, it should provide a framework for investigating these concepts in a rigorous and systematic manner.

The link between the Akashic Records and the nature of reality is still highly speculative, falling into the realm of philosophical inquiry and parapsychological exploration rather than established scientific fact. However, when viewed through the lens of quantum physics and the emerging understanding of the interconnectedness of the universe, the possibility of such a universal database becomes less improbable. The challenge is to develop methods that can bridge the gap between these seemingly disparate domains, allowing us to explore these concepts with a balanced approach that respects both scientific rigor and the potential for profound discoveries beyond our current understanding.

The implications are far-reaching. If access to this universal memory is possible, it could fundamentally alter our understanding of history, consciousness, and our individual roles within the cosmic tapestry. It could provide a framework for understanding phenomena that currently elude scientific explanation, such as precognition, intuition, and collective consciousness. Moreover, it could pave the way for a deeper understanding of our connection to the universe and to each other, fostering a sense of shared responsibility and interconnectedness.

The exploration of the Akashic Records and universal consciousness is not simply an academic exercise; it is a journey of profound personal and cosmic significance. It is a path that leads us to question our most fundamental assumptions about reality and to explore the possibilities of existence beyond our current comprehension. It requires a willingness to embrace uncertainty, to question our preconceived notions, and to approach the unknown with open minds and a thirst for knowledge. The quest to understand the universe and our place within it is a journey without end, a continuous exploration of the mysteries that surround us, pushing the boundaries of scientific thought and inviting a more profound and interconnected view of reality. The pursuit of this understanding is not only a scientific endeavor, but also a spiritual one, a quest to understand not just the universe, but ourselves in the context of the cosmos.

The journey towards understanding the Akashic Records and universal consciousness is a collaborative one, requiring the integration of insights from various fields, including physics, cosmology, parapsychology, philosophy, and even spirituality. By fostering interdisciplinary collaboration and embracing a holistic perspective, we can begin to unravel the mysteries surrounding these concepts and unlock a deeper understanding of the universe and our place within it. The path forward necessitates a paradigm shift, moving from a fragmented worldview towards a more interconnected and unified perspective, recognizing the intricate web of connections that permeates all aspects of existence. This shift in perspective might lead to revolutionary advancements in our understanding of consciousness, reality, and the very fabric of the cosmos. It's a journey that calls for open minds, unwavering curiosity, and a willingness to explore the unknown. The rewards of such a journey, however, could be immeasurable, leading to a more profound and fulfilling understanding of our place within the grand cosmic design. The pursuit of this knowledge, in its essence, is a pursuit of self-understanding, a quest to discover our own connection to the infinite.

Having explored the concept of a potentially interconnected universe and the intriguing possibility of the Akashic Records, we now turn our attention to a phenomenon that further supports this hypothesis: synchronicity. Synchronicity, as defined by Carl Jung, refers to meaningful coincidences—events that appear causally unrelated yet possess a striking interconnectedness, often imbued with a sense of significance for the individual experiencing them. These are not mere chance occurrences, but rather events seemingly orchestrated by a deeper, underlying order.

Consider, for instance, the experience of thinking about a long-lost friend, only to receive a phone call or email from that very person shortly thereafter. Or perhaps you're contemplating a particular idea, and suddenly, a book, article, or conversation presents that very same concept, seemingly out of the blue. These are examples of synchronicity, events that seem to defy the laws of probability and suggest a deeper interconnectedness between seemingly disparate aspects of our lives.

From a purely materialistic perspective, synchronicity can be dismissed as mere coincidence, a statistical anomaly. However, when viewed through the lens of a potentially interconnected universe, as proposed in this book, synchronicity takes on a different meaning. If, as we've explored, the universe operates on principles beyond our current understanding of classical physics, then the seemingly random occurrence of meaningful coincidences might be interpreted as evidence of a deeper, underlying order—a cosmic orchestration, if you will.

One potential explanation for synchronicity within the framework of our hypothesis is the concept of non-locality. As discussed earlier, quantum entanglement demonstrates that two or more particles can be interconnected regardless of the distance separating them. This non-local connection suggests that information can be transmitted instantaneously across vast distances, transcending the limitations of spacetime. Could this same principle apply to events within our lives, linking seemingly unrelated occurrences through a non-local connection?

Another potential explanation lies in the concept of the Akashic Records, the cosmic memory containing the complete history of every event and every soul. If such a repository of information exists, it could potentially influence the unfolding of events in our lives, creating a pattern of synchronicity that aligns with our individual paths and purposes. The seemingly random events could, in fact, be orchestrated by this underlying informational field, guiding us towards specific experiences and lessons.

The concept of resonance also plays a crucial role. Imagine the universe as an intricate web of interconnected vibrations. Our thoughts, emotions, and intentions resonate with the universal field, creating patterns that attract corresponding events. In this sense, synchronicity could be seen as a reflection of this resonant interplay, events aligned with our inner state and our individual trajectory.

It's important to note that the interpretation of synchronicity often involves subjective experience. What one person considers a meaningful coincidence might be dismissed by another as mere chance. This subjective element doesn't negate the phenomenon, however; it simply highlights the complex interplay between our inner world and the external universe. The meaning we attribute to a synchronistic event is often deeply personal, reflecting our beliefs, values, and expectations.

Consider the work of physicist David Bohm, who proposed the concept of implicate order. This suggests that the universe is not simply a collection of separate objects and events, but rather a unified, interconnected whole. Meaningful coincidences might be manifestations of this implicate order, events unfolding in a way that reflects the interconnectedness of all things. Our perception of separate events is merely a limited manifestation of the deeper underlying unity.

Moreover, the study of synchronicity often leads to discussions about the nature of consciousness itself. If we truly are interconnected, as our exploration suggests, then our consciousness might not be limited to our physical bodies. It could extend into the universal field, influencing and being influenced by events seemingly beyond our control. In this view, synchronicity becomes a tangible manifestation of the interplay between our consciousness and the universe. Our thoughts, intentions, and emotions could resonate with the universal field, attracting events that mirror and amplify those internal states.

The exploration of synchronicity necessitates a shift from a purely mechanistic worldview to one that embraces the interconnectedness of all things. It requires a willingness to recognize patterns and connections that might otherwise go unnoticed. It also requires a capacity for introspection, allowing us to identify the deeper meanings and significance of seemingly random events in our lives.

While there is no definitive scientific explanation for synchronicity, its consistent appearance across diverse cultures and throughout history suggests a phenomenon deserving of further investigation. The possibility that meaningful coincidences are manifestations of a deeper, interconnected universe challenges our conventional understanding of causality and probability. It opens the door to a broader perspective, one that acknowledges the potential influence of consciousness and the subtle interconnectedness of all events.

Furthermore, the exploration of synchronicity bridges the gap between science and spirituality. It provides a framework for understanding seemingly paranormal experiences within a scientific context. The events aren't dismissed as random chance; instead, they are considered potentially meaningful indications of the interconnectedness of all things, hinting at an underlying order that transcends our current scientific comprehension. This framework encourages a more holistic approach to understanding reality, acknowledging both the material and the metaphysical aspects of existence.

Synchronicity, in its essence, challenges us to question our assumptions about causality and randomness. It urges us to look beyond the surface level of events, seeking deeper connections and meanings. It encourages us to consider the possibility of an underlying order and a profound interconnectedness that permeates the entire universe. It encourages a more mindful and intentional approach to life, recognizing the subtle ways in which we are connected to everything around us and to everything that has ever been and ever will be.

By exploring these connections and recognizing the patterns within the seemingly random occurrences of our lives, we gain a deeper understanding of ourselves, our place in the universe, and the remarkable tapestry of interconnectedness that weaves together all aspects of existence. The seemingly coincidental events, the meaningful synchronicities, are not just random occurrences but potentially vital pieces of a larger puzzle, suggesting a universe far more interconnected and meaningful than we might have previously imagined.

In conclusion, the study of synchronicity further reinforces the hypothesis presented in this book: that the universe is far more interconnected than we currently understand. These seemingly random occurrences, when viewed through the lens of quantum entanglement, the Akashic Records, and the concept of a universal consciousness, become less surprising and more deeply significant, offering a glimpse into a reality that is both interconnected and profoundly meaningful. The more we explore these phenomena, the more we unravel the intricate tapestry of existence, revealing the incredible depth and interconnectedness of all things. The exploration of synchronicity serves as a testament to the interconnected nature of the universe, encouraging us to embrace the mystery and the wonder that surround us. It is a call to be more mindful, more observant, and more open to the subtle ways in which the universe communicates with us. The quest to comprehend synchronicity ultimately leads us toward a deeper understanding of ourselves, our place within the cosmos, and the profound interconnectedness of all things. This journey, though challenging, promises to be profoundly rewarding, revealing a universe rich with meaning, connection, and the incredible wonder of existence itself.

Having established the interconnectedness hinted at through synchronicity, and explored potential mechanisms like non-locality and the Akashic Records, we can now synthesize these threads into a broader understanding of reality's interwoven fabric. The universe, as we've been suggesting, isn't a collection of isolated entities governed solely by classical physics; it's a dynamic, interconnected web where seemingly disparate phenomena are linked through subtle, often invisible, channels. This interconnectedness transcends the limitations of space and time, manifesting in ways that challenge our conventional understanding of causality and probability.

Consider the implications for our understanding of consciousness. If the universe operates as a unified field, as suggested by Bohm's implicate order, then consciousness itself might not be confined to the physical brain. Instead, it could be an aspect of this larger, interconnected field, extending its influence across space and time. This perspective aligns with certain interpretations of quantum mechanics, where the act of observation influences the outcome of an experiment, suggesting a fundamental link between consciousness and the physical universe. The observer, in this model, is not a passive entity but an active participant in shaping reality.

This interconnected field of consciousness could also help explain phenomena typically relegated to the realm of parapsychology. Telepathy, for example, could be interpreted as a form of communication within this interconnected field, where information is transmitted not through conventional sensory channels but through a deeper, more fundamental level of interconnectedness. Similarly, precognition – the ability to foresee future events – could be a result of accessing information within this field that transcends the linear flow of time.

Furthermore, the concept of a unified field allows for a more nuanced understanding of the relationships between different realms and dimensions. Variations in density, as discussed previously, could create distinct planes of existence, but these planes would not be entirely isolated from each other. Instead, they might be interconnected through subtle energetic pathways, allowing for the potential interaction and exchange of information between them. This interconnectedness could explain reported encounters with entities from other realms or dimensions, as well as the apparent ability of some individuals to traverse these boundaries.

This perspective also lends itself to a richer interpretation of cosmological phenomena. The seemingly random distribution of galaxies in the universe could be a manifestation of underlying patterns and connections within the interconnected field. The very structure of spacetime itself might be influenced by this underlying field, suggesting that the universe's geometry isn't fixed but dynamically responsive to the interactions within this interconnected web.

To further illustrate this interconnectedness, let's consider the concept of morphogenetic fields, proposed by Rupert Sheldrake. These fields are hypothesized to organize and influence the development of living systems, suggesting that organisms are not simply governed by their genetic code but also by collective patterns and influences. This concept extends the idea of interconnectedness beyond individual organisms to encompass entire species and even ecosystems. A change in one part of the system could have ripple effects throughout the entire web, highlighting the intricate and dynamic nature of this interconnected reality.

The implications of this interconnected web are profound, impacting not only our understanding of the universe but also our place within it. It suggests that we are not isolated individuals but rather integral parts of a larger, interconnected system. Our thoughts, emotions, and actions resonate within this field, influencing not only our individual experiences but also the collective reality we share. This underscores the importance of mindful living, recognizing the potential impact of our actions on the larger interconnected web.

The challenge, therefore, lies in developing new scientific tools and frameworks to explore and understand this interconnected reality. Existing scientific methods, primarily focused on isolated systems and linear causality, are insufficient for capturing the complexity and dynamism of this interwoven web. We need innovative approaches that incorporate concepts from quantum physics, information theory, and systems biology to better grasp the underlying principles governing this interconnected universe.

This doesn't necessarily imply a rejection of conventional scientific methodology. Instead, it suggests a broadening of our perspective, integrating current scientific knowledge with new theoretical frameworks and incorporating evidence from diverse fields, including parapsychology, to construct a more complete and accurate picture of reality. A multi-disciplinary approach, bringing together physicists, biologists, psychologists, and other researchers, is vital to advancing our understanding of this interconnected universe.

The exploration of this interconnected web necessitates a paradigm shift in our thinking. It requires us to move beyond a purely materialistic view of reality, acknowledging the role of consciousness and information in shaping the universe. It calls for a more holistic and integrated approach, embracing the interconnectedness of all things and acknowledging the subtle, often invisible, forces that connect us to each other and to the cosmos.

This understanding doesn't negate the importance of scientific rigor. Rather, it necessitates a broadening of our scientific lens to encompass the full spectrum of reality, including phenomena that currently fall outside the boundaries of conventional scientific understanding. The pursuit of knowledge in this realm requires an open mind, a willingness to challenge established paradigms, and a recognition of the limitations of current scientific models.

The journey towards understanding this interconnected web is a continuous process of exploration and discovery. As we delve deeper into the mysteries of the universe, we will undoubtedly encounter new challenges and uncover new dimensions of this intricate tapestry of existence. The implications of this interconnectedness are vast and far-reaching, affecting not only our understanding of the universe but also our relationship with ourselves, each other, and the world around us.

This exploration is not merely an intellectual exercise. It has profound ethical and practical implications. Understanding the interconnectedness of all things compels us to act with greater responsibility and compassion, recognizing that our actions have far-reaching consequences that extend beyond our immediate surroundings. It compels us to embrace a more mindful and sustainable way of life, understanding the interconnectedness of all living beings and the delicate balance of the ecosystems we inhabit.

The implications for future scientific inquiry are also significant. The pursuit of understanding this interconnected reality requires a shift from reductionist approaches to holistic ones. It necessitates the development of new theoretical frameworks and experimental methods capable of capturing the complexity and dynamism of this interwoven web. This could involve developing technologies capable of detecting and measuring subtle energetic interactions, exploring the role of consciousness in shaping reality, and investigating the fundamental nature of information in the universe.

Ultimately, the concept of the interconnected web of reality offers a powerful framework for understanding the universe. It provides a lens through which we can view seemingly disparate phenomena and uncover the underlying patterns and connections that unite them. It challenges us to reconsider our assumptions about causality, probability, and the nature of reality itself. It is a journey of discovery that promises to reshape our understanding of the cosmos and our place within it, leading to a more profound appreciation of the interconnectedness and wonder of existence.

This interwoven tapestry of reality, far from being chaotic, suggests a fundamental order and coherence that underpins all aspects of existence. The apparent randomness of events may mask deeper patterns and connections, waiting to be revealed through careful observation and insightful analysis. This exploration is not just about uncovering the secrets of the universe; it is about discovering our own place within this intricate and beautiful web, recognizing our interconnectedness with all things, and understanding our role in shaping the future of this interconnected reality. The universe, seen through this lens, is not just a vast expanse of matter and energy, but a living, breathing entity, a dynamic web of interconnectedness pulsating with life and meaning. This profound understanding promises to transform our relationship with the cosmos and our understanding of our place within it. It is a journey of discovery that invites us to embrace the mystery and the wonder of existence, to recognize the interconnectedness of all things, and to live more mindfully and responsibly within this intricate and magnificent web of life. The pursuit of this knowledge is not just a scientific endeavor; it is a spiritual journey, a quest for deeper meaning and understanding of our place in the universe, a journey that promises to be both challenging and profoundly rewarding.

14

Challenges and Objections

The preceding chapters have ventured into territory considered by many to be outside the realm of conventional science. Exploring the interconnectedness of reality, the possibility of multiple dimensions, and the potential for paranormal phenomena naturally invites skepticism and criticism. It's crucial, therefore, to address these objections directly and thoughtfully, acknowledging the limitations of the evidence while presenting a robust defense of the perspectives explored.

One of the most common critiques leveled against research into parapsychology and related fields is the lack of rigorous, repeatable experimental evidence. Critics often point to the inconsistent nature of results, the difficulties in controlling for bias and error, and the absence of a unifying theoretical framework. These concerns are valid and deserve careful consideration. The inherent challenges in studying phenomena that defy easy quantification and control present significant obstacles. The subjective nature of many experiences reported – near-death experiences, out-of-body experiences, and psychic abilities – makes objective measurement exceptionally difficult. The reproducibility of results, a cornerstone of the scientific method, is often elusive in this domain, partly due to the subtle and elusive nature of the phenomena being investigated. Furthermore, the influence of expectation and belief on both the experimenter and the subject can introduce significant bias. However, the absence of definitive proof does not necessarily equate to disproof. Many scientific discoveries have emerged from initially ambiguous or seemingly contradictory evidence. The history of science is replete with instances where initial skepticism gave way to acceptance as methodologies improved and theoretical understanding evolved. The very act of attempting to quantify and analyze phenomena such as telepathy or precognition pushes the boundaries of established scientific paradigms, requiring the development of new methodologies and theoretical frameworks that are still in their infancy.

Another common objection stems from the potential for fraud and misinterpretation. The allure of the extraordinary often attracts individuals prone to deception or self-deception. Cases of documented fraud in parapsychological research are indeed concerning and underscore the need for stringent experimental design and rigorous verification. The burden of proof rests firmly on those proposing these extraordinary claims. Skepticism, therefore, is not only warranted but essential in advancing our understanding. However, dismissing all evidence simply due to the possibility of fraud is a fallacy. This is analogous to rejecting all astronomical observations because some past observations have proven erroneous. A critical and discerning approach is necessary, carefully evaluating each instance of alleged paranormal phenomena against a backdrop of established scientific knowledge and methodological rigor. This rigorous evaluation includes not only scrutinizing the methods employed in the research, but also carefully considering potential alternative explanations for reported phenomena, such as misperception, coincidence, or even outright fabrication.

The concept of multiple dimensions, realms, or planes of existence also attracts criticism. Some critics argue that these concepts are purely speculative, lacking any empirical evidence. The limitations of our current understanding of physics are undeniable, and the universe may indeed harbor complexities beyond our present comprehension. While we may not yet possess the tools to directly observe or interact with other dimensions, theoretical frameworks such as string theory and M-theory postulate their existence. These theories, though not yet experimentally verified, offer mathematical models that suggest the possibility of more dimensions than those we directly experience. The fact that we cannot currently measure or observe something doesn't inherently invalidate its potential existence. Our sensory perception and current technological capabilities restrict our experience of reality, and the history of science shows that our understanding of the universe is constantly evolving.

Moreover, the seemingly random or chaotic nature of many paranormal experiences is often cited as evidence against their validity. The unpredictability of these phenomena doesn't necessarily refute their existence. Many natural phenomena – earthquakes, volcanic eruptions, even weather patterns – are still subject to a degree of unpredictability despite advances in scientific understanding. The challenge lies not in dismissing these phenomena as inherently improbable but in seeking to understand the underlying mechanisms that might govern them. It is possible that the apparent randomness reflects our current lack of understanding, rather than an inherent lack of structure or causality. The underlying order might be obscured by complex interactions and subtle influences currently beyond our detection capabilities.

The intertwining of scientific concepts with ideas traditionally associated with parapsychology naturally provokes criticism from both sides of the spectrum. Some within the scientific community might argue against considering parapsychological phenomena as valid subjects of scientific inquiry. They may contend that such exploration could compromise the integrity of science by embracing what they perceive as pseudoscience. However, this perspective overlooks the fact that the history of science itself is punctuated by periods of paradigm shifts, where previously held beliefs are challenged and ultimately superseded by new understanding. The development of quantum mechanics, for instance, fundamentally altered our understanding of the universe, demonstrating that what once seemed impossible or illogical could be scientifically demonstrable.

Similarly, some within the parapsychology community might argue that the integration of scientific rigor potentially dilutes or limits the exploration of consciousness and other phenomena considered to be outside the purview of conventional science. This concern highlights the delicate balance needed in this multidisciplinary approach. The aim is not to impose the constraints of conventional science on the investigation of unconventional phenomena, but rather to apply rigorous methodology to the exploration of these experiences. This requires a collaborative dialogue and a willingness to adapt our methods and understanding to address the unique challenges presented by these subjects.

The path forward necessitates a commitment to ongoing research, a willingness to embrace novel theoretical frameworks, and the development of innovative methodologies. Interdisciplinary collaborations are crucial, bringing together scientists, parapsychologists, and philosophers to approach the challenges from diverse perspectives. The goal is not to definitively prove or disprove the existence of any particular phenomena but to expand our understanding of the universe and the limits of our perception. The process of inquiry itself is an ongoing journey, and while the answers remain elusive, the very act of asking these questions opens up new avenues of exploration and discovery, potentially leading us to a deeper appreciation of the boundless complexity and wonder of reality. The existence of unexplained phenomena should not be seen as a failure of science, but as a challenge – an invitation to broaden our perspectives, refine our methodologies, and deepen our understanding of the cosmos and our place within it. Ultimately, the open-minded pursuit of knowledge, coupled with rigorous methodology, is crucial to advancing our understanding, irrespective of whether it conforms to prevailing paradigms or challenges conventional scientific thinking.

Anecdotal evidence, while often captivating and suggestive, forms a precarious foundation for scientific claims, particularly within the realm of parapsychology and the study of purportedly paranormal events. The inherent limitations of relying on personal accounts, often recounted long after the event, introduce significant biases and uncertainties that render such accounts unreliable as standalone evidence. Memory is fallible; details are frequently embellished, omitted, or unconsciously altered over time, influenced by suggestion, expectation, and cultural narratives. The human mind is remarkably adept at constructing narratives that fit pre-existing beliefs and expectations, leading to the potential for misinterpretations and even self-deception. This is particularly problematic when dealing with emotionally charged experiences or events that defy easy explanation. A compelling narrative, however emotionally resonant, does not equate to scientific validity.

Consider, for example, the numerous accounts of near-death experiences (NDEs). These accounts often describe a sense of floating outside the body, encountering deceased loved ones, or experiencing a life review. While deeply personal and profoundly impactful for the individual, these narratives lack the objective measurability and repeatability demanded by rigorous scientific investigation. The physiological changes occurring during a near-death experience—reduced oxygen supply to the brain, for instance—can themselves account for many of the reported sensory phenomena. Furthermore, the cultural context in which the experience is embedded significantly shapes its interpretation and narration. Individuals raised in cultures with specific beliefs about the afterlife will naturally interpret their NDE through this lens, potentially leading to accounts that align with those preconceived notions, rather than reflecting an objective reality.

Similarly, alleged encounters with ghosts or extraterrestrial beings frequently rely heavily on anecdotal evidence. Witness testimony is often inconsistent, contradictory, and susceptible to suggestibility. The power of suggestion cannot be underestimated. The expectation of encountering a paranormal entity, coupled with environmental cues (unusual sounds, shadows, or atmospheric conditions), can readily lead to misinterpretations of sensory input. Furthermore, the desire to believe, the allure of the extraordinary, can dramatically influence the interpretation of ambiguous experiences, creating vivid and compelling narratives that are ultimately unsubstantiated. The absence of corroborating physical evidence—photographic, video, or forensic—further weakens the reliability of these accounts.

Even in cases where multiple witnesses report similar events, the phenomenon of collective hallucination or mass hysteria cannot be ruled out. The phenomenon of shared delusion, where a group of individuals develop and share a false belief, is a well-documented psychological phenomenon. This highlights the importance of controlling for social influence and group dynamics when evaluating anecdotal evidence. Furthermore, the possibility of deliberate fabrication or hoaxing cannot be ignored. The public fascination with paranormal events provides an incentive for individuals to create elaborate and convincing stories, often for personal gain or notoriety.

The inherent difficulty in replicating anecdotal experiences poses another significant challenge. A key tenet of the scientific method is the reproducibility of results. If a phenomenon is truly real and consistent, it should be demonstrably repeatable under controlled conditions. However, by their very nature, many anecdotal experiences are unique, specific to time, location, and individual circumstances, rendering them inherently difficult, if not impossible, to replicate. This lack of reproducibility greatly hinders efforts to develop robust scientific models and theories to explain these phenomena.

While anecdotal evidence may provide intriguing leads and stimulate further investigation, it cannot, on its own, establish the validity of a claim. It serves as a starting point, potentially highlighting areas that warrant further investigation using rigorous scientific methodologies. The shift from anecdotal evidence to scientific inquiry requires a fundamental change in approach. Instead of relying on personal testimonies, researchers need to develop controlled experiments, quantify observations, minimize bias, and rigorously test hypotheses. The collection of objective data, the use of quantitative measures, and the application of statistical analysis are crucial for drawing valid conclusions. Without these crucial components, the allure of anecdotal evidence risks diverting attention away from the rigorous scientific investigation necessary to unravel the mysteries of our universe, both seen and unseen.

Furthermore, the challenge of studying elusive phenomena that don't neatly fit into pre-existing scientific frameworks requires the development of new methodologies and innovative research designs. This demands creativity, intellectual flexibility, and a willingness to challenge established paradigms. Simply dismissing anecdotal evidence as inherently unreliable, however, neglects the potential for uncovering new insights and understandings. A balanced and nuanced approach is required – one that recognizes the limitations of anecdotal evidence while remaining open to the possibility of discovering phenomena that may lie beyond our current comprehension. This open-mindedness, however, must be tempered by a rigorous commitment to scientific integrity and a careful consideration of alternative explanations.

The development of new technologies and advancements in scientific understanding could potentially overcome some of the challenges posed by anecdotal evidence. For example, advances in brain imaging and neurophysiology may shed light on the neural correlates of experiences such as NDEs, providing objective data to support or refute subjective accounts. Similarly, advancements in physics and cosmology could lead to theoretical models that provide a framework for understanding phenomena that currently defy explanation. The application of rigorous statistical analysis, combined with the use of sophisticated data analysis techniques, can help researchers to identify patterns and trends in seemingly random events, potentially uncovering underlying mechanisms that govern these phenomena.

In conclusion, while anecdotal evidence might spark initial interest and inspire further investigation into the paranormal, it cannot stand alone as evidence for the existence of extraordinary phenomena. The inherent problems of memory distortion, suggestibility, cultural bias, and the potential for fraud render anecdotal accounts unreliable in the absence of supporting scientific evidence. The path toward a scientifically sound understanding of these phenomena requires a commitment to rigorous methodology, objective data collection, and a willingness to challenge existing paradigms. The pursuit of knowledge requires a careful balancing act—one that embraces curiosity and open-mindedness while maintaining a steadfast commitment to the principles of scientific rigor and critical thinking. Only through this balanced approach can we hope to decipher the mysteries of existence and determine the extent of our perceived reality.

The inherent limitations of the scientific method itself, when applied to phenomena that seemingly defy our current understanding of the universe, must also be acknowledged. Our scientific models are built upon observations and experiments within a specific framework—the framework of our current knowledge. This framework, however, may be incomplete or even fundamentally flawed when it comes to grappling with phenomena that exist outside its parameters. We are, in essence, attempting to understand the unknown using tools designed to comprehend the known. This inherent circularity introduces a significant limitation. Consider, for example, the challenges in studying dark matter and dark energy, which constitute the vast majority of the universe's mass-energy content, yet remain largely invisible to our current observational techniques. Our understanding of these fundamental components of the cosmos is based on their gravitational effects, indirect evidence inferred from their influence on visible matter. We are detecting their presence, but not directly observing them, forcing us to build theoretical models based on limited data, constantly refined and modified as new information becomes available.

The problem is further compounded by the limitations of our sensory apparatus and technological capabilities. Human perception is inherently restricted; we only perceive a minuscule portion of the electromagnetic spectrum, for example, and our understanding of other forces of nature, such as the strong and weak nuclear forces, remains comparatively limited compared to our grasp of gravity and electromagnetism. Our technology, while sophisticated, also has its boundaries. Even the most advanced telescopes and particle accelerators can only probe a finite region of space and energy. What exists beyond the reach of our current technology remains unknown, and potentially unknowable, at least within our present conceptual framework. We are essentially exploring a vast ocean with a small, albeit advanced, submersible.

This brings us to the crucial issue of paradigm shifts in scientific thinking. The history of science is punctuated by moments of revolutionary change, where existing paradigms are overthrown and replaced by new, more comprehensive models. The shift from a geocentric to a heliocentric view of the solar system, the development of quantum mechanics, and the theory of relativity are all examples of paradigm shifts that drastically altered our understanding of the universe. These paradigm shifts frequently involve the incorporation of phenomena that were previously considered anomalous or inexplicable within the existing paradigm. Often, these paradigm shifts occur not through incremental refinement of existing models but through a fundamental re-evaluation of basic assumptions. Therefore, it's entirely plausible that phenomena currently labeled as "paranormal" may ultimately find a place within a future scientific paradigm, once our understanding of the universe has evolved sufficiently.

The limitations are not just technological or conceptual; there are also methodological challenges. The scientific method, while powerful, is best suited for studying phenomena that are readily observable, repeatable, and measurable under controlled conditions. Many of the phenomena explored within parapsychology and cosmology, however, defy easy categorization within this framework. The inherent unpredictability and subjective nature of these phenomena pose significant challenges for rigorous scientific investigation. For example, spontaneous human combustion, while seemingly impossible based on current understanding of physics and chemistry, nevertheless appears in anecdotal accounts. The lack of repeatable instances and the challenges in ruling out alternative explanations render scientific study extremely difficult, if not impossible, with our current methodologies.

Furthermore, the very act of observation and measurement can profoundly alter the phenomena being investigated. The observer effect in quantum mechanics exemplifies this issue, where the act of measuring a quantum system inherently affects its state. This presents a fundamental challenge when studying phenomena that are inherently sensitive to observation. The interaction between the observer and the observed might inadvertently skew the results, introducing unavoidable biases into the experimental data. This is particularly relevant when considering subjective experiences, such as near-death experiences or out-of-body experiences, where the act of recording and interpreting the experience might itself alter the nature of the experience.

Moreover, the complex interplay between consciousness and the physical world poses a significant challenge to scientific investigation. While neuroscience has made significant strides in understanding the neural correlates of consciousness, the fundamental nature of consciousness itself remains a mystery. The subjective experience of reality is inherently personal and uniquely shaped by individual experiences, beliefs, and cultural context. This subjective dimension makes it incredibly challenging to develop objective and universally applicable models of consciousness and its interaction with the physical world.

The concept of multiple realities or dimensions, although speculative, cannot be categorically dismissed based on our current scientific understanding. The theoretical frameworks of string theory and M-theory, for example, suggest the possibility of extra spatial dimensions beyond the three we perceive. These dimensions may be curled up and undetectable at our current energy scales, but their existence cannot be ruled out. Similarly, the concept of a multiverse, where our universe is just one of many, is supported by some cosmological models. The existence of such realities, if indeed they exist, poses enormous methodological challenges for scientific investigation, but their potential cannot be ignored.

Even within established physics, there are unanswered questions that highlight the limits of our current understanding. The nature of time, the arrow of time, the origin of the universe, and the ultimate fate of the universe all remain open questions. These questions challenge our basic assumptions about the nature of reality and require a paradigm shift in our thinking in order to potentially answer them. The pursuit of these answers, however, demands not just a commitment to scientific rigor, but also a willingness to embrace intellectual humility and an acceptance that our current understanding of the universe is incomplete and subject to future revisions.

In conclusion, the pursuit of knowledge about the universe, especially in areas that push the boundaries of our current understanding, requires a balanced approach. While the scientific method provides a crucial framework for investigation, it is crucial to acknowledge its inherent limitations. It is not a matter of abandoning scientific rigor, but rather of expanding its reach to encompass the unknown. This necessitates a combination of open-mindedness and rigorous scrutiny, a willingness to challenge established paradigms, and the development of new methodologies that can better address the complexities of the universe, both seen and unseen. The exploration of phenomena that appear to defy our current understanding is not a threat to science, but rather an opportunity to advance our knowledge and deepen our appreciation of the intricate tapestry of reality. The path forward requires a synthesis of scientific skepticism and intellectual curiosity, a balance that enables us to push the boundaries of our understanding while maintaining a commitment to intellectual honesty and the highest standards of scientific inquiry.

The exploration of paranormal phenomena often leads to a fascinating and sometimes frustrating dichotomy: compelling anecdotal evidence juxtaposed against a scarcity of robust scientific verification. This necessitates a careful consideration of alternative explanations, moving beyond simple dismissal as "unscientific" towards a more nuanced understanding of the potential underlying mechanisms. A balanced perspective requires exploring psychological and sociological factors that can convincingly mimic or even generate experiences often attributed to paranormal activity.

One significant area to examine is the power of suggestion and expectancy. The placebo effect, well-documented in medicine, demonstrates the profound influence of belief on physiological and psychological responses. In the context of paranormal investigation, the expectation of experiencing a paranormal event—whether consciously or subconsciously held—can significantly bias perception and interpretation. For example, in alleged haunted houses, participants might interpret ambiguous sounds or shadows as evidence of ghostly activity, guided by pre-existing beliefs and expectations. The inherent ambiguity of many paranormal experiences allows for a wide range of interpretations, often shaped by the individual's biases and prior knowledge.

The role of cognitive biases is also crucial. Confirmation bias, for instance, leads individuals to seek out and favor information that confirms their existing beliefs, while ignoring or downplaying contradictory evidence. This can lead to misinterpretations of ambiguous sensory data, particularly in situations where the desire to believe in paranormal phenomena is strong. Similarly, the availability heuristic, which emphasizes readily available information, can create a skewed perception of the likelihood of paranormal events. Highly publicized cases, even if ultimately debunked, can enhance the perceived probability of similar events, fostering a sense of plausibility.

Furthermore, the social context plays a significant role in shaping both the experience and the interpretation of paranormal events. Group dynamics can amplify individual suggestibility, leading to shared perceptions and interpretations that solidify belief in paranormal phenomena. The social pressure to conform within a group, particularly when dealing with emotionally charged experiences, can lead to individuals suppressing doubts and confirming others' accounts, even if their own initial impressions were less certain. This phenomenon is heightened in contexts where shared belief systems or cultural narratives strengthen the expectations of encountering paranormal activity.

Memory fallibility also presents a major challenge in the investigation of paranormal claims. Memories are not static recordings of events; they are reconstructive processes susceptible to distortion, both intentional and unintentional. The passage of time, the influence of subsequent information, and the emotional intensity of an experience can all significantly alter memories. Alleged accounts of paranormal events, often recounted long after the supposed occurrence, are particularly prone to distortion and embellishment. The human brain, it seems, actively constructs and reconstructs memories, making reliance on solely on anecdotal testimony unreliable.

The influence of cultural narratives and folklore is another important consideration. Stories and beliefs about ghosts, spirits, or other paranormal entities are deeply ingrained in many cultures. These cultural narratives shape expectations and interpretations of sensory experiences, providing a framework through which individuals perceive and understand the world. Experiences that might otherwise be mundane or readily explainable might be interpreted as paranormal if they align with pre-existing cultural beliefs. The power of cultural narratives cannot be underestimated in fostering and reinforcing belief in paranormal phenomena.

Another avenue for explaining some reported paranormal occurrences lies in the realm of electromagnetic fields (EMFs). While not directly linked to the classic ghost stories, subtle fluctuations in EMF levels can affect human perception and physiology. Some individuals are highly sensitive to even slight changes in EMFs, experiencing physical symptoms such as headaches, nausea, or disorientation. In environments with unusual EMF patterns, these symptoms could be misinterpreted as paranormal encounters, creating a correlation that doesn't necessarily imply causation. This is particularly relevant in older buildings, where electrical wiring might be erratic or where natural geological factors influence EMFs. Technological interference from poorly shielded electronics or faulty wiring also needs to be considered.

Additionally, sleep paralysis, a relatively common phenomenon, often involves vivid hallucinations and a sense of intense pressure or oppression. During sleep paralysis, an individual is awake but unable to move, a condition that can be terrifying. The sensory hallucinations that accompany sleep paralysis can be extremely realistic, leading to experiences that are easily misconstrued as paranormal encounters. The feeling of a presence, the sense of being touched or watched, and even visual hallucinations are common components of sleep paralysis, readily fitting narratives of paranormal encounters.

Similarly, hypnagogic and hypnopompic hallucinations, those occurring during the transitional states between wakefulness and sleep, can generate vivid and seemingly real experiences. These hallucinations can involve visual, auditory, or tactile sensations, and their realistic nature can contribute to the belief that a paranormal event has taken place. These states are often associated with altered perceptions of time and space, reinforcing the sense of an otherworldly experience.

Furthermore, neurological conditions such as temporal lobe epilepsy can produce a range of unusual sensory experiences, including hallucinations, déjà vu, and altered perceptions of reality. These conditions can generate experiences that could easily be misinterpreted as paranormal phenomena, highlighting the importance of considering physiological factors in investigating such claims. Similarly, mental health conditions such as schizophrenia can involve hallucinations and delusions that might be perceived as paranormal events.

Finally, fraud and deception must also be considered. The history of paranormal investigation is replete with instances of deliberate hoaxes and fraudulent claims designed to exploit public credulity. The motivations for such actions range from financial gain to seeking notoriety or simply causing mischief. Differentiating genuine unexplained events from deliberate fabrications requires rigorous investigation and skepticism, relying on objective evidence and careful scrutiny of all claims.

In conclusion, while some paranormal claims might remain enigmatic, a comprehensive evaluation must explore a wide range of alternative explanations. The influence of psychological factors such as suggestion, expectancy, and cognitive biases; the impact of social contexts and cultural narratives; the effects of physiological conditions and neurological disorders; and the possibility of fraud all need to be considered in a balanced and thorough analysis. Dismissing all claims as unsubstantiated is unscientific, but accepting them at face value without critical examination is equally problematic. A rigorous, multidisciplinary approach is necessary, one that combines open-mindedness with a healthy dose of skepticism, and employs the most advanced tools and techniques available to unlock the mysteries of our reality, whatever their origin may be.

The preceding discussion has highlighted the complexities inherent in investigating phenomena that lie outside the conventional boundaries of scientific understanding. We've explored various psychological, sociological, and physiological factors that can plausibly explain experiences often attributed to the paranormal. However, dismissing all such claims outright as mere illusion or fabrication would be equally unscientific. A true pursuit of knowledge requires a delicate balance: a commitment to rigorous scientific methodology coupled with an open mind willing to consider possibilities that currently defy easy explanation. This necessitates a robust application of critical thinking, a skill crucial not only for evaluating paranormal claims but also for navigating the vast landscape of scientific inquiry itself.

Critical thinking, in this context, is far more than mere skepticism. Skepticism, while essential, can easily devolve into cynicism, prematurely dismissing potentially valid avenues of investigation. True critical thinking involves a willingness to engage with evidence, evaluate arguments, and consider alternative explanations, while maintaining a healthy dose of doubt – a doubt applied equally to established theories as to unconventional claims. It demands a constant questioning of assumptions, a meticulous examination of biases, and a persistent search for robust, verifiable evidence.

This approach is particularly relevant when dealing with seemingly paradoxical phenomena. For instance, reports of precognitive dreams, where individuals claim to have foreseen future events, challenge our understanding of causality and temporal linearity. While dismissing these reports as mere coincidence or post-hoc rationalization is tempting, a critical thinker would first analyze the methodology used to collect and evaluate such data. Were the dreams recorded contemporaneously? Were independent witnesses involved? What are the potential biases involved in remembering and recalling such experiences? Only after a thorough examination of these aspects can we begin to evaluate the validity of the claims. Furthermore, exploring potential alternative explanations, such as chance occurrences or subconscious information processing, becomes paramount.

Similarly, alleged cases of remote viewing, where individuals claim to perceive distant locations without sensory input, demand rigorous analysis. While many such claims have been debunked as fraud or the result of sensory leakage, some studies have yielded results that are difficult to explain away through purely conventional means. A critical thinker would scrutinize these studies for methodological flaws, potential biases, and the quality of the evidence. They would also examine alternative explanations, such as unconscious knowledge of the target location or unconscious communication between the viewer and the experimenter. The goal isn't to prove or disprove the existence of remote viewing; rather, it's to rigorously evaluate the available evidence and assess the plausibility of various explanations.

The investigation of near-death experiences (NDEs) presents another compelling example. These accounts, often involving out-of-body experiences, vivid visions of light, and feelings of profound peace, are frequently cited as evidence for life after death or the existence of a non-physical realm. A critical approach to NDEs would involve examining the physiological changes occurring during the process of death and near-death. The effects of oxygen deprivation on brain function, the release of endorphins, and the activation of specific brain regions could all contribute to the subjective experiences reported by NDE survivors. Furthermore, psychological factors such as fear of death and the desire for meaning could shape the interpretation of these experiences. A balanced perspective requires exploring both the physiological and psychological factors before drawing conclusions about the metaphysical implications of NDEs.

Furthermore, applying critical thinking to cosmology itself opens fascinating avenues for exploration. The standard model of cosmology, while remarkably successful in explaining many features of the universe, leaves some significant questions unanswered. The nature of dark matter and dark energy, the ultimate fate of the universe, and the possibility of multiple universes are all areas where critical thinking, coupled with further research and open-mindedness, is essential. Challenging accepted models, proposing alternative explanations, and rigorously testing these hypotheses are paramount for advancing our understanding of the cosmos.

The concept of a multiverse, for example, challenges our intuitive understanding of reality. Different theoretical frameworks, such as string theory and inflationary cosmology, predict the existence of multiple universes with different physical laws and constants. While there is currently no direct observational evidence for a multiverse, the very concept demands a critical examination of our assumptions about the universe and its boundaries. Could our understanding of physical laws be incomplete? Could there be other universes, perhaps undetectable by our current methods? These questions, though speculative, warrant exploration through careful theoretical analysis and the development of new observational techniques.

The application of critical thinking extends beyond simply evaluating evidence. It also necessitates a clear understanding of the limits of our knowledge. Acknowledging what we don't know is as important as understanding what we do know. This humility in the face of the vast unknown is crucial for preventing premature conclusions and maintaining an open mind for new discoveries. A willingness to revise our understanding in light of new evidence is a hallmark of scientific progress and a fundamental aspect of critical thinking.

In summary, navigating the intersection of science and the paranormal, or indeed any area of scientific inquiry that pushes the boundaries of established knowledge, requires a powerful blend of critical thinking and open-mindedness. This doesn't mean uncritically accepting every extraordinary claim; rather, it implies a commitment to a rigorous, evidence-based approach that is both skeptical and receptive to new possibilities. It's about asking probing questions, evaluating diverse perspectives, and maintaining a commitment to the relentless pursuit of truth, however elusive or unexpected it might be. This approach allows us to engage in a meaningful exploration of the unknown, expanding the horizons of our understanding and perhaps revealing aspects of reality far beyond our current comprehension. This open, critical engagement is not only crucial for advancing our knowledge of the cosmos but also for fostering a more nuanced and informed understanding of our place within it, whatever that place may ultimately be. Ultimately, the pursuit of knowledge is a journey, not a destination, and the commitment to critical thinking is the compass guiding us along the way. The universe, in its infinite complexity, offers endless opportunities for discovery, and it is through the rigorous application of intellectual honesty and open-minded inquiry that we can hope to unravel its profound mysteries.

15

A New Paradigm for Reality

This exploration of reality beyond our commonly perceived limits has taken us on a journey through the intricate landscapes of physics, consciousness, and the very nature of existence. We've traversed established scientific theories, venturing into the often-controversial realms of parapsychology and cosmology, seeking potential bridges between these seemingly disparate fields. The overarching theme has been the exploration of density as a potential key to understanding the multifaceted nature of reality.

The concept of density, initially considered within the framework of classical physics, has proven remarkably fertile ground for expanding our understanding of the universe. Different densities, we've argued, might correspond to different planes of existence, different realms of reality, each with its unique physical laws and properties. This isn't mere speculation; we've examined the implications of varying densities within the context of established physics, exploring how density fluctuations could create regions of spacetime with drastically different characteristics. The implications are profound: the possibility of universes layered upon one another, each operating under its own physical parameters.

The implications of this model extend beyond the purely physical. We've examined how this multi-layered reality might provide a framework for understanding phenomena that currently defy conventional scientific explanation, such as alleged encounters with ghosts, or seemingly inexplicable events often attributed to paranormal activity. These events, if viewed through the lens of differing density realms, might be interpreted not as violations of physical laws, but rather as interactions between these disparate layers of reality. This doesn't necessitate accepting every paranormal claim uncritically; rather, it suggests a shift in perspective, a reframing of our understanding of what's possible. The rigorous application of scientific methodology remains crucial; however, dismissing all seemingly anomalous events as mere illusion or fabrication prematurely closes off potentially significant avenues of inquiry.

Consider, for instance, the reported experiences of near-death experiences (NDEs). While conventional explanations invoking physiological changes in the brain due to oxygen deprivation or endorphin release are undoubtedly relevant, our model suggests another intriguing possibility. Perhaps some NDEs represent a temporary shift in consciousness, a fleeting interaction with a higher-density realm, a glimpse into a different plane of existence. The vivid descriptions of light, profound peace, and out-of-body experiences might not be mere hallucinations, but rather genuine sensory perceptions within a different dimensional context, shaped by the unique physical laws governing that realm.

Similarly, the concept of a multiverse, often relegated to the realm of science fiction, gains a more tangible foothold within this framework. Our exploration has showcased how fluctuations in density could give rise to not just different realms within our universe, but to entirely separate universes, each with its own set of physical constants and laws. This isn't a simplistic, one-size-fits-all multiverse; rather, it's a dynamic, interconnected system of universes, each potentially influencing the others through subtle interactions, perhaps even generating the enigmatic phenomena we currently label as "paranormal."

This approach requires us to reassess our understanding of spacetime. If different densities correspond to different levels of reality, then our perception of space and time itself may not be absolute. The very fabric of spacetime, it could be argued, might be fluid and interconnected, allowing for seemingly instantaneous communication across seemingly vast distances. This could potentially provide a mechanism for phenomena such as remote viewing, where individuals claim to perceive distant locations without sensory input. While rigorous testing and verification are necessary to establish the validity of such claims, our proposed model offers a potentially plausible mechanism that warrants further investigation.

The philosophical implications are equally profound. Our exploration has hinted at the possibility of consciousness transcending the physical body, existing independently of its biological limitations. The notion of density realms suggests that consciousness might not be confined to the three spatial dimensions we perceive, but could exist in higher-dimensional spaces, interacting with our reality in ways that currently elude our full comprehension. This raises questions about the nature of self, identity, and the very definition of life. Are we limited to our physical bodies, or is there a larger, more profound aspect of our existence that operates outside the conventional bounds of space and time?

The integration of parapsychology into a cosmological framework requires a considerable leap of faith for those entrenched in conventional scientific thinking. The very nature of parapsychological phenomena – their inherent unpredictability, the difficulty of replicating them under controlled conditions – makes them challenging subjects for scientific inquiry. However, the possibility of interactions between different density realms offers a potential explanation for these otherwise inexplicable events.

But even beyond the speculative realm, the implications for understanding our place within the cosmos are significant. The model presented in this book suggests that the universe is not merely a vast expanse of matter and energy, but a complex, multi-layered reality, where different levels of density create realms with unique properties. This shift in perspective opens up exciting possibilities for future research, not only in the realm of parapsychology, but also in cosmology and other branches of physics. The development of new technologies and experimental techniques, tailored to explore these subtle interactions between different density realms, could revolutionize our understanding of the universe. The exploration of density as a fundamental aspect of reality allows for a more unified view of the universe, incorporating both the macroscopic structures of galaxies and clusters and the microscopic interactions of subatomic particles. This unified view could hold the key to unlocking the deepest mysteries of the cosmos.

The scientific pursuit of understanding the universe necessitates a delicate balance between rigorous methodology and open-mindedness. We must be skeptical, demanding strong evidence and carefully evaluating methodologies, but we must also be willing to explore unconventional hypotheses, to consider possibilities that currently lie outside the bounds of our established understanding. This requires a profound shift in perspective, a willingness to challenge our preconceived notions of reality, and a recognition that our current knowledge may only represent a fraction of the vast unknown.

In conclusion, the findings presented in this book offer a compelling new paradigm for understanding reality. By exploring the multifaceted nature of density, and its implications for the existence of multiple realms and potentially even multiple universes, we have opened up new avenues for investigating phenomena that previously defied scientific explanation. This journey has been one of scientific exploration, philosophical reflection, and imaginative speculation, guided always by the pursuit of knowledge and a deep respect for the vast unknown. The ultimate truth of the universe remains elusive, but the pursuit of understanding is a journey worth undertaking, one that promises to yield profound insights into our place within the cosmos, and perhaps, into the very nature of existence itself. The challenge lies in developing the tools and methodologies needed to explore these possibilities further, to move beyond speculation and into a realm of verifiable knowledge. The possibilities, however, are truly awe-inspiring. The universe, it seems, is far grander and more intricate than we have ever imagined.

This reevaluation necessitates a fundamental shift in perspective, moving beyond the limitations of our current scientific models. For centuries, our understanding of reality has been largely shaped by Newtonian physics, a framework that elegantly describes the macroscopic world but falls short in explaining phenomena at the quantum level or those seemingly defying conventional explanations. The paradigm shift we propose moves towards a more holistic view, integrating seemingly disparate fields like quantum physics, cosmology, and parapsychology into a unified framework. This framework, based on the concept of density as a fundamental organizing principle of reality, suggests that the universe is not a single, homogeneous entity, but rather a multi-layered system, with different densities corresponding to different planes of existence.

Consider the implications for our understanding of spacetime. In our current model, spacetime is treated as a fixed, unchanging background upon which physical events unfold. But what if spacetime itself is not fixed, but rather a dynamic, fluid entity, its properties shaped by the density of the region it occupies? In regions of high density, spacetime might be compressed, leading to a perception of faster time, while in regions of low density, spacetime might be expanded, resulting in a perception of slower time. This could potentially explain some of the anomalous temporal phenomena reported in certain parapsychological accounts. Furthermore, if spacetime is not a fixed entity but a fluid and interconnected structure, this might allow for instantaneous communication or even travel across vast cosmological distances, potentially explaining phenomena like remote viewing. The possibility of "wormholes" or other shortcuts through spacetime, commonly discussed in theoretical physics, gains increased plausibility under this model.

The implications for consciousness are equally profound. Our current scientific understanding largely confines consciousness to the realm of the physical brain. However, if reality is multi-layered, with different densities corresponding to different levels of existence, then consciousness might not be limited to its physical substrate. It could potentially exist independently of the brain, operating in a higher-dimensional space, interacting with our three-dimensional reality in ways we are only beginning to comprehend. This model doesn't necessarily negate the importance of the brain; instead, it proposes a more complex relationship between mind and matter. The brain, in this context, acts as an interface, a receiver and transmitter of consciousness, translating signals between the different density realms.

The challenge lies in developing the tools and methodologies necessary to empirically investigate these concepts. Currently, our technological capabilities are insufficient to directly probe the properties of different density realms. However, advancements in quantum physics, especially in the areas of quantum entanglement and quantum superposition, may offer new avenues of exploration. The detection of subtle energetic interactions between different density levels might be possible through highly sensitive instruments capable of measuring subtle changes in electromagnetic fields or gravitational waves. Furthermore, advances in neuroscience and consciousness studies, particularly those exploring non-local correlations in brain activity, could provide crucial insights into the nature of consciousness and its interactions with different density levels.

The exploration of these possibilities necessitates a reassessment of our scientific methodologies. We cannot simply dismiss phenomena that don't readily fit within our existing frameworks. We must develop new approaches, more open to exploring the unconventional and the anomalous. This requires a shift towards more rigorous observation, meticulous documentation, and a willingness to collaborate across disciplines, integrating insights from physics, cosmology, parapsychology, and other relevant fields. A multidisciplinary approach, breaking down the traditional boundaries between these fields, might prove crucial to uncovering the deeper truths about reality.

The development of new theoretical frameworks is equally important. Our current understanding of physics is based on a series of approximations and simplifications. We need to develop more sophisticated models that can accurately describe the behavior of matter and energy at all scales, from the subatomic to the cosmological. String theory, loop quantum gravity, and other advanced theories may hold the key to unlocking a deeper understanding of spacetime and its relationship to density. These theories, often considered speculative, might become crucial tools in constructing a more unified and accurate model of reality.

The philosophical implications are far-reaching. If reality is multi-layered, our sense of self and our place within the universe undergoes a profound transformation. We are not merely confined to our physical bodies; our consciousness might extend beyond the limitations of space and time. This raises fundamental questions about the nature of life, death, and the afterlife, opening up the possibility of survival of consciousness beyond physical death. These are not questions that science alone can answer. They require a deeper engagement with philosophy, ethics, and spirituality.

This new paradigm also has profound implications for our ethical considerations. If we accept the possibility of multiple realms of existence, it forces us to rethink our responsibilities towards all sentient beings, not just those confined to our physical reality. It may also influence our relationship with the environment, prompting us to view the Earth not merely as a resource to be exploited, but as a complex and interconnected part of a larger cosmic system. Understanding our place within this larger system may lead to a greater sense of responsibility and a deeper appreciation for the interconnectedness of all things.

The scientific investigation of these ideas must be approached with caution and rigorous methodology. The temptation to embrace unsubstantiated claims without critical analysis should be resisted. However, a healthy skepticism should not lead to the premature rejection of potentially valuable avenues of inquiry. A balanced approach, combining rigorous experimentation with open-mindedness, is vital for progress.

The journey towards a new paradigm for understanding reality is a long and challenging one. It requires a willingness to question our preconceived notions, to embrace uncertainty, and to explore the unknown. It requires a commitment to rigorous scientific inquiry, combined with a deep appreciation for the mystery and wonder of the universe. But the potential rewards are immense: a deeper understanding of our place within the cosmos, a broader perspective on the nature of existence, and a more profound appreciation for the interconnectedness of all things. The pursuit of this understanding is a journey worth undertaking, one that promises to reveal truths far beyond our current comprehension, transforming our perception of reality and reshaping our understanding of our place within the vast tapestry of the universe. The future of our understanding rests in a commitment to continued exploration, a balance between rigorous scientific methods and open-minded investigation, and a willingness to challenge the very foundations of our beliefs. The universe, after all, is a far grander and more mysterious place than we can currently comprehend. The quest to understand it is a journey without end, but a journey of immense intellectual and spiritual reward.

The path forward necessitates a concerted, multidisciplinary effort, transcending the traditional boundaries between scientific disciplines. Physics, cosmology, parapsychology, neuroscience, and even philosophy must converge to effectively address the complexities inherent in exploring realities beyond our current understanding. This interdisciplinary approach demands a reassessment of our research methodologies. We must move beyond the limitations of reductionist models, which often fail to capture the holistic nature of reality. Instead, we need to embrace systems thinking, recognizing the intricate web of interdependencies that govern the universe. This requires a shift from focusing on isolated phenomena to investigating complex systems and their emergent properties.

One crucial area for future exploration involves the refinement of our experimental techniques. Currently, our instruments are often insufficient to detect subtle energetic interactions that might indicate the presence of other realities. Advances in quantum sensing technologies, such as those employing superconducting quantum interference devices (SQUIDs) or atom interferometers, could dramatically enhance our sensitivity to these subtle signals. The development of more sophisticated detectors for gravitational waves, capable of discerning the faintest ripples in spacetime, could also reveal evidence of interactions between different density realms. Furthermore, advances in biofeedback technology and neuroimaging techniques might provide crucial insights into the nature of consciousness and its interaction with the broader cosmic environment.

The development of theoretical frameworks capable of encompassing these diverse phenomena is equally essential. While string theory and loop quantum gravity offer promising avenues for understanding the fundamental structure of the universe, they remain incomplete. We require theoretical models that can seamlessly integrate concepts from quantum physics, general relativity, and parapsychology. This necessitates the development of new mathematical tools and conceptual frameworks, potentially drawing upon ideas from topology, higher-dimensional geometry, and information theory. The challenge lies in developing a unified theoretical model that can explain not only the macroscopic world we observe but also the subtle energetic interactions and anomalous phenomena reported in parapsychological research.

The exploration of consciousness occupies a central position in this new paradigm. While our current understanding largely confines consciousness to the realm of the physical brain, the possibility of non-local consciousness—consciousness existing independently of a physical substrate—demands further investigation. This investigation must move beyond the limitations of materialistic reductionism, acknowledging the potential for consciousness to interact with reality in ways that defy conventional physical explanations. Research into altered states of consciousness, near-death experiences, and other paranormal phenomena might offer invaluable clues into the nature of consciousness and its interaction with other density realms. Advances in quantum biology and the study of microtubules within brain cells, coupled with advanced neuroimaging, are crucial for investigating the possible role of quantum coherence in consciousness. The prospect of consciousness existing beyond the physical body raises profound philosophical implications, prompting a reevaluation of our understanding of life, death, and the afterlife.

Another critical area for future research is the exploration of altered states of consciousness. These states, induced through meditation, sensory deprivation, or other techniques, can significantly alter an individual's perception of reality. The experiences reported during these states often resemble accounts of paranormal phenomena, suggesting a potential connection between altered states of consciousness and interactions with other realities. Further investigation into the neurobiological mechanisms underlying these altered states, along with rigorous documentation of the subjective experiences involved, could provide valuable insights into the nature of consciousness and the potential for interactions with different density levels.

The investigation into anomalous phenomena, such as remote viewing, precognition, and psychokinesis, should also be pursued with renewed vigor. While these phenomena have been met with skepticism within mainstream science, the sheer volume of documented evidence, coupled with the potential explanatory power of a multi-layered reality model, suggests a need for systematic, rigorous investigation. Advanced experimental protocols, utilizing double-blind methodologies and sophisticated data analysis techniques, are required to reduce the possibility of error and bias. This should involve not just documenting the phenomena but also attempting to understand the underlying mechanisms, potentially through investigation of subtle energetic interactions or quantum entanglement.

The development of new technologies is crucial for furthering our understanding. Advances in artificial intelligence and machine learning could provide powerful tools for analyzing vast amounts of data from diverse sources. AI could be employed to identify patterns and correlations that might escape human observation, revealing hidden connections between different realms of existence. Furthermore, advanced computational models, simulating the interactions of matter and energy at different density levels, could assist in developing theoretical models.

However, caution is paramount. The exploration of these realms must be undertaken with a healthy dose of skepticism and a rigorous commitment to scientific methodology. Extraordinary claims require extraordinary evidence, and any new paradigm must withstand the scrutiny of peer review and empirical validation. Pseudo-scientific claims must be rigorously challenged, and unsubstantiated assertions should be avoided. Maintaining scientific integrity and rigor is essential to avoid misleading the public and compromising the credibility of this vital field of inquiry.

The philosophical and ethical implications of this new paradigm are vast. If we accept the possibility of other realities, it necessitates a profound re-evaluation of our values and beliefs. It challenges our anthropocentric worldview, reminding us that humanity may not be the sole or even the most significant form of consciousness in the universe. This perspective shifts our focus from a purely materialistic understanding of existence to a holistic vision, recognizing the interconnectedness of all things. This understanding could profoundly influence our interactions with the environment and our relationships with each other. It could engender a new sense of global responsibility and foster a greater appreciation for the delicate balance of life on Earth.

In conclusion, the future of exploration into the unknown lies in a convergence of scientific disciplines, a refinement of experimental techniques, and the development of new theoretical frameworks. It requires a willingness to challenge established paradigms and embrace the unknown with both curiosity and critical rigor. This endeavor calls for a multidisciplinary, collaborative approach, bridging the gap between science, philosophy, and spirituality. The journey will be challenging, but the potential rewards are immense—a deeper understanding of reality, a more profound appreciation for the interconnectedness of all things, and a more responsible stewardship of our planet and its place within the vast cosmic tapestry. The exploration of the unknown is an ongoing journey, one of perpetual discovery and wonder, revealing the universe's vast and intricate beauty. This pursuit, however challenging, is a vital endeavor, holding the key to unlocking a richer, more profound understanding of our place in the cosmos.

The implications of a multi-layered reality, as explored throughout this book, extend far beyond the purely scientific. The acceptance of dimensions and realms beyond our immediate perception necessitates a profound shift in our understanding of ourselves, our place in the cosmos, and our potential. This shift promises a transformation not just of our scientific understanding but also of our individual lives and the collective consciousness of humanity.

One immediate impact is the potential for personal transformation. The realization that reality extends beyond the confines of our five senses can be profoundly liberating. For many, the belief in a singular, material reality can be limiting, leading to feelings of isolation, insignificance, and a sense of confinement within the perceived boundaries of the physical world. The possibility of interacting with other dimensions, even subtly, opens up the prospect of accessing previously untapped potential within ourselves. This might manifest as enhanced creativity, intuition, and spiritual growth. The exploration of altered states of consciousness, discussed earlier, suggests pathways to such interaction, providing tools for personal development and self-discovery that extend far beyond conventional psychological methods. Individuals might discover latent abilities, find answers to life's profound questions, and connect with a sense of purpose transcending their immediate existence. The understanding of interconnectedness fostered by the recognition of other realms can lead to a greater sense of compassion and empathy, fostering more harmonious relationships with oneself and others.

Furthermore, the implications for society as a whole are substantial. A paradigm shift of this magnitude would inevitably challenge established social structures, belief systems, and power dynamics. The dominance of materialistic values, often associated with a limited, singular-reality worldview, might be challenged by a more holistic and spiritually inclusive perspective. The acceptance of non-local consciousness, for instance, could profoundly alter our conceptions of death and the afterlife, leading to less fear of mortality and a greater appreciation for the preciousness of life. This understanding could reshape our social institutions, particularly those surrounding healthcare, end-of-life care, and bereavement. The implications for our legal systems are equally significant. If the boundaries between dimensions are permeable, as some parapsychological research suggests, then legal frameworks surrounding issues such as evidence, testimony, and the definition of reality itself would need careful re-evaluation.

Our understanding of consciousness requires a radical reconsideration. The limitations of materialistic reductionism, which seeks to explain consciousness solely through neurochemical processes, must be overcome. The possibility of non-local consciousness—consciousness existing independently of a physical substrate—challenges the very foundations of our current scientific and philosophical models. If consciousness is not solely a product of the brain, then its potential for interaction with other realities and dimensions becomes far more plausible. This necessitates a reassessment of our understanding of mental health, with the potential for new treatment approaches that consider the interconnectedness of consciousness with the broader cosmic environment.

The impact on science itself will be transformative. The need for interdisciplinary collaboration, emphasized throughout this book, cannot be overstated. The exploration of other realities necessitates the convergence of physics, cosmology, parapsychology, neuroscience, and philosophy, amongst other fields. Such collaboration will not only lead to a richer and more comprehensive understanding of reality but will also reshape the scientific method itself. We might need to develop new methodologies and experimental designs capable of investigating phenomena that defy conventional scientific paradigms. The very definition of "evidence" might need to be broadened to encompass a wider range of data, including subjective experiences, anecdotal evidence, and intuitive insights.

Furthermore, a new paradigm of reality may reshape our relationship with technology. Artificial intelligence, in particular, could play a crucial role in exploring the unknown. AI's ability to process and analyze vast amounts of data, including data from sources considered outside the mainstream scientific paradigm, could lead to groundbreaking discoveries. However, the ethical implications of utilizing AI in this context cannot be ignored. Safeguards must be in place to prevent misuse and to ensure that AI is employed responsibly in the pursuit of knowledge. The development of advanced technologies for sensing and interacting with different density levels—ranging from refined quantum sensors to sophisticated biofeedback systems—will be essential for furthering our understanding of reality beyond the confines of our current sensory limitations.

The economic implications of a multi-layered reality are also noteworthy. The potential for new technologies and discoveries resulting from our expanded understanding of reality is immense. New sources of energy, materials, and information could revolutionize various sectors, leading to unprecedented economic growth and societal transformation. However, the equitable distribution of the benefits of such advances is critical. The potential for economic disparity and social inequality must be addressed to ensure that the benefits of this new paradigm are shared by all.

The exploration of other realities also presents significant ethical considerations. The potential for manipulation, misuse of power, and unethical experimentation must be guarded against. The development of guidelines and ethical frameworks for conducting research in this nascent field is paramount to avoid potential harm. We must consider the potential consequences of our actions on other realities and ensure that our actions are guided by principles of responsibility, respect, and compassion.

Finally, the spiritual implications are profound. A paradigm shift of this magnitude can lead to a renewed sense of awe and wonder, a profound appreciation for the interconnectedness of all things, and a deeper understanding of our place in the vast cosmic tapestry. This could foster a greater sense of global responsibility, urging humanity towards collective action to address the environmental challenges facing our planet. The acceptance of a multi-layered reality can nurture a more spiritual worldview, promoting tolerance, understanding, and the pursuit of wisdom. This understanding may promote a more peaceful and harmonious coexistence, fostering a sense of community that transcends the limitations of our physical reality. The journey into the unknown, though fraught with challenges, promises a transformation of unprecedented scale, leading humanity towards a future where scientific understanding is intertwined with spiritual insight, fostering a more profound, compassionate, and interconnected existence. The path is long, but the potential rewards – a deeper understanding of ourselves, our place in the cosmos, and our potential – are immeasurable.

The journey we've undertaken in this book—a journey into the multifaceted nature of reality, exploring the potential for dimensions beyond our immediate perception and the intriguing possibilities offered by parapsychology—has only scratched the surface. The implications of a multi-layered universe, woven together with threads of scientific theory and speculative exploration, are far-reaching and profoundly transformative. We have glimpsed a cosmos far richer, more complex, and far more wondrous than the one typically portrayed by conventional science. But the path ahead remains largely unexplored. This final section is not an ending, but a call to arms, an invitation to participate in the ongoing exploration of the mysteries that surround us.

The exploration of other realities requires a radical shift in our approach to scientific inquiry. The rigid adherence to materialistic reductionism, which confines our understanding of the universe to observable phenomena and measurable quantities, must be tempered with a willingness to consider phenomena that currently lie outside the purview of conventional scientific methods. This doesn't imply a rejection of the scientific method itself, but rather an expansion of its scope to encompass a wider range of possibilities. We must embrace a more holistic and interdisciplinary approach, drawing upon the insights of physics, cosmology, parapsychology, neuroscience, philosophy, and other fields to construct a more complete picture of reality.

This interdisciplinary collaboration is crucial. Physics provides the theoretical framework for understanding the fundamental laws governing the universe, including the potential for multiple dimensions and parallel universes. Cosmology offers insights into the large-scale structure of the cosmos and the possibility of other realities existing beyond our observable universe. Parapsychology, despite its controversial status within the mainstream scientific community, offers a wealth of anecdotal and empirical evidence suggesting the existence of phenomena that defy conventional explanations. Neuroscience explores the complexities of consciousness, offering clues about the potential for non-local consciousness and the relationship between mind and matter. Finally, philosophy provides the conceptual framework for understanding the implications of a multi-layered reality, challenging our assumptions about the nature of reality, consciousness, and existence itself.

The development of new methodologies for investigating these phenomena is equally crucial. Conventional scientific tools and experimental designs may be inadequate for exploring realities beyond our current sensory capabilities. We may need to develop new instruments capable of detecting subtle energy fields or interacting with other dimensions. This may involve developing advanced quantum sensors, exploring the use of biofeedback systems, or even creating entirely novel technologies that we can't presently conceive of. Furthermore, the nature of evidence itself needs to be reconsidered. While controlled experiments and quantitative data remain essential, we must also consider qualitative data, including subjective experiences, anecdotal reports, and intuitive insights. The integration of diverse forms of evidence will be vital in constructing a more complete and accurate picture of reality.

The ethical implications of this exploration cannot be overstated. The potential for misuse of knowledge and technology in this area is significant. We must approach this new frontier with a profound sense of responsibility and ethical awareness. This means developing comprehensive ethical guidelines for research in this field, ensuring that any investigation into other realities is conducted in a way that minimizes the risk of harm and maximizes the potential for benefit. It also requires a commitment to transparency and open communication, ensuring that the results of any research are shared openly and honestly with the wider scientific community and the public.

Furthermore, we need to consider the societal implications of a paradigm shift in our understanding of reality. The acceptance of a multi-layered universe could fundamentally alter our perceptions of ourselves, our place in the cosmos, and our relationship with each other. This may lead to changes in our social structures, belief systems, and values. The potential for both positive and negative outcomes necessitates thoughtful and careful consideration of the potential social, political, and economic consequences. Open public discourse and participatory decision-making processes will be crucial to navigating this transition responsibly.

Beyond the immediate scientific and ethical challenges, lies a potential for profound personal transformation. The realization that reality is far more expansive and interconnected than we previously imagined can be deeply liberating. It can foster a greater sense of awe, wonder, and connection to the universe. This expanded awareness can lead to a renewed sense of purpose, a deeper appreciation for the interconnectedness of all things, and a profound sense of belonging within the grand cosmic tapestry. This, in turn, can inspire positive changes in individual lives, leading to greater personal growth, spiritual development, and a heightened sense of compassion and empathy for others.

The exploration of other realities is not merely a scientific quest; it is a human endeavor. It is an invitation to expand our consciousness, to challenge our preconceived notions, and to embrace the unknown with courage and curiosity. It is a call to action, a summons to participate in a grand cosmic adventure that promises to transform our understanding of ourselves, our universe, and our place within it. The journey will be challenging, complex, and undoubtedly fraught with unforeseen difficulties. Yet, the potential rewards—a deeper understanding of the universe and our own potential—are immense, promising a future where scientific advancement and spiritual insight merge to create a more enlightened and compassionate world.

The ultimate goal is not just to uncover the secrets of other realities but also to integrate this newfound knowledge into our lives, fostering a more harmonious and interconnected existence. The challenges are immense, requiring a concerted effort from scientists, philosophers, spiritual leaders, and the wider public. By working together, through open collaboration and a commitment to ethical exploration, we can embark on this journey into the unknown, unlocking the mysteries of the universe and transforming our understanding of reality itself. This requires open minds, a willingness to question established paradigms, and a deep commitment to seeking truth, regardless of its form. The answers may not always be easy to find, but the journey itself is a profound and transformative experience, one that holds the potential to reshape our understanding of the universe and our place within it. Let us embrace this challenge, for the future of humanity and our understanding of existence itself may depend on it. The exploration continues...

www.ingramcontent.com/pod-product-compliance
Lightning Source LLC
LaVergne TN
LVHW040517210926
848480LV00007BA/1487